SOVIET SCIENCE
1917-1970

Part I
ACADEMY OF SCIENCES OF THE USSR

Edited by
PAUL K. URBAN, ANDREW I. LEBED

The Scarecrow Press, Inc.
Metuchen, New Jersey, USA
1971

Assistant Editors:
(English)

ELLIOTT, J. B.   KRAFT, E. G.

Compiled at
THE INSTITUTE FOR THE STUDY OF THE USSR
Munich · Germany

BY

MELINSKAYA, S. I., RESEARCH ASSISTANT

Library of Congress Catalog Card No.: 77—161562
ISBN 0—8108—0440—9

Copyright 1971, by
The Scarecrow Press, Inc.

CONTENTS

Introduction

    Preface i
    Transliteration System iii
    Abbreviations and Terms iv

CHAPTER I

    Outline History of the USSR Academy of Sciences

        Pre-Soviet Period (1725-1917)   1
        The Soviet Period (1917-1970) 17

CHAPTER II

    Heads of the USSR Academy of Sciences

        Presidium of the USSR
        Academy of Sciences 53

        Departments of the USSR
        Academy of Sciences 56

        Institutions of the USSR Academy of
        Sciences' Presidium 63

        Institutions of the USSR
        Academy of Sciences 66

        Branches and Bases of the USSR
        Academy of Sciences 127

        Siberian Division of the USSR
        Academy of Sciences 133

        Notes 147

        Publications of the USSR
        Academy of Sciences 178

CHAPTER III

Members of the USSR Academy of Sciences

Full, Corresponding and Honorary Members of the USSR Academy of Sciences 196

Members of the Russian Academy of Sciences 1725-1917 266

Foreign Members of the USSR Academy of Sciences 301

Supplements

International Organizations 303

National Committees and Associations of Soviet Scientists 308

Gold Medals and Prizes Awarded by the USSR Academy of Sciences 311

Data Received After Press Deadline 316

Sources 321

PREFACE

In publishing this guide, the Institute for the Study of the USSR aims to provide as detailed a review as possible of the development of the USSR Academy of Sciences from 1917 to 1970.

In December 1962 the Institute published a short guide to the Academy of Sciences, but since that time there have been major changes in the Academy, particularly in 1963 when, on the basis of the CPSU Central Committee and USSR Council of Ministers' resolution of 11 April on "Measures Governing Improvements in the Work of the USSR Academy of Sciences and the Union-Republic Academies of Sciences," new Statutes of the USSR Academy of Sciences were drafted and adopted on 1 July 1963. Inter alia, the new Statutes greatly increased the USSR Academy of Sciences' jurisdiction over the union-republic academies. In view of these changes, the Institute feels that the present guide will prove a timely aid to Free World researchers studying the development of science in the USSR. The guide is the forerunner in a series of works on the development and organization of science in the USSR.

The structure and status of the USSR Academy of Sciences differ essentially from that of Western academies of sciences. The prime difference is that all its Presidium's decisions on scientific plans, new members and the establishment of new research institutions require the approval of the USSR Council of Ministers. In the West the academies of sciences are more or less free corporations of scientists and scholars and have only a small number of research establishments, since most Western research work is conducted at universities or by private industry, whereas Soviet research is almost entirely conducted by the establishments of the USSR Academy of Sciences. Under the new 1963 Statutes the USSR Academy of Sciences even controls the work of research establishments which are not under its direct jurisdiction, as well as that of the research establishments subordinate to the republican academies of sciences. In his report to the General Assembly of the USSR Academy of Sciences on 14-15 May 1963, Academy President M.V. Keldysh said:

> "In order to improve the USSR Academy of Sciences' liaison with the union-republic academies of sciences and to centralize the management of science, all matters connected with the development of social and natural sciences in our country will be considered and decided by the Presidium of the USSR Academy of Sciences. The work of the union-republic academies of sciences will be directed by both the Presidium of the USSR Academy of Sciences and the corresponding union-republic council of ministers."

The guide contains three sections and several supplements. The first section is an outline history of the USSR Academy of Sciences. The second examines the structure and organization of the Academy and lists the directors of its various establishments. Entries requiring explanation or amplification are numbered and annotated at the end of the section. The third section contains various lists, including registers of full, corresponding and foreign members of the USSR Academy of Sciences.

INSTITUTE TRANSLITERATION SYSTEM

Russian-English

А, а	- A, a		Р, р	- R, r
Б, б	- B, b		С, с	- S, s
В, в	- V, v		Т, т	- T, t
Г, г	- G, g		У, у	- U, u
Д, д	- D, d		Ф, ф	- F, f
Е, е	- E, e (ye)*		Х, х	- Kh, kh
Ё, ё	- Yo, yo		Ц, ц	- Ts, ts
Ж, ж	- Zh, zh		Ч, ч	- Ch, ch
З, з	- Z, z		Ш, ш	- Sh, sh
И, и	- I, i		Щ, щ	- Shch, shch
Й, й	- Y, y		Ъ, ъ	- ' (b')
К, к	- K, k		Ы, ы	- Y, y
Л, л	- L, l		Ь, ь	- ' (n')
М, м	- M, m		Э, э	- E, e
Н, н	- N, n		Ю, ю	- Yu, yu
О, о	- O, o		Я, я	- Ya, ya
П, п	- P, p			

Adjectival endings

ый	—	yy
ий	—	iy
ые	—	yye
ие	—	iye

* "ye" is used initially, after vowels (а, е, ё, и, о, у, ы, э, ю, я), and after ъ and ь ; elsewhere it is "e."

General: Where a family or place name (Gorky, Kiev) or the title of a publication *(Izvestia)* have a generally accepted or standardized transliteration, it will be used in our publications.

Abbreviations and Terms

Acad of Sci	Academy of Sciences
Assoc	Association, associate
Asst	Assistant
Bd	Board
Centr	Central
Comt	Committee
Corresp	Corresponding
Dept	Department
Ed	Editor, edition
Electr	Electricity, electrical
Eng	Engineering, engineer
Exec	Executive
Gen	General
Govt	Government, governmental
Hon	Honorary
Ind	Industry, industrial
Inst	Institute
Int	International
Lit	Literature, literary
Mech	Mechanics, mechanical
Med	Medicine, medical
Min	Minister
Prof	Professor
Publ	Published
Sci	Science, scientific
Secr	Secretary
Soc	Society
Tech	Technical, technics
Technol	Technology, technological
Telemech	Telemechanics
Vil	Village
Vol	Volume

Chapter I

OUTLINE HISTORY OF THE USSR ACADEMY OF SCIENCES

1. Pre-Soviet Period

Russia's Academy of Sciences was founded in 1725 and was closely linked with the reforms that Peter I instituted in the first quarter of the 18th Century.

As a result of his 1697-98 foreign tour, Peter I resolved to introduce science into Russia with the aim of modernizing the army and navy, developing industry and promoting trade and crafts. By the late 17th Century Russia had two higher educational establishments with quite extensive programs - the Greco-Latin-Slavic Academy in Kiev and a similar academy, modeled on the former, in Moscow. These academies were higher theological schools with a scholastic curriculum, although the Kiev Academy did also dabble in the new, experimental science. In 1714 Peter I founded the Kunstkammer and Library as independent institutions. The Naval Academy and Engineering School were founded in 1715 and 1719 as educational establishments.

In the early 18th Century Russia lacked scientists of its own, but there were many enlightened persons who backed Peter I's reformist plans. On several occasions Peter I had instructed them to "form an academy," but he neglected to define its nature or tasks. It was not until 1719, after he had visited and become a member of the Paris Academy, that Peter came around to the idea of patterning a Russian Academy on the one in Paris. With the end of the war against Sweden in 1721 he perfected his plan of establishing a Russian Academy of Sciences with an affiliated university and high school.

On 24 January 1724 Peter I approved the draft for the Academy of Sciences, promulgated by the Senate on

28 January in the form of a nominal decree. Preparations to found the academy began that same year with the recruiting of staff and negotiations with foreign scholars and scientists who were to take the place of academicians. But Peter I did not live to see the opening of the Academy of Sciences; he died on 28 January 1725.

The establishment of the Academy of Sciences was the last link in Peter I's chain of reforms. Earlier he had organized the first Kamchatka Expedition to settle the question of whether Asia were connected with America. After Peter's death the organization of the Academy of Sciences took almost an entire year until the first scientific assembly of the academicians was held on 2 November 1725 (Old Style). The Latin minutes of this assembly have been preserved. On 27 December 1725 this was followed by the ceremonial opening of the Academy of Sciences, in public assembly at Shafirov's house in the Petersburg Quarter, since the Academy building on Vasiliy Ostrov was not yet completed.

Originally the institution was named the Russian Academy, for a short while the Petersburg Academy, then until 1917 the Imperial Academy, and from 1917 to 1925 again the Russian Academy. For a considerable time its members were foreigners, including such prominent scientists as the mathematicians J. Hermann, N. and D. Bernoulli and Leonard Euler and the historian Gerhard-Friedrich Miller. In the period from 1725 to 1740 only one of the Academy's 44 full members was Russian - the mathematician V.Ye. Adadurov, who came to the Academy in 1733. In order to train Russian scientists as academicians the Academy had its own university (which existed until the 1760's) and high school (which operated until 1805). Between 1741 and 1751 the Academy gained seven more Russian scientists, including the encyclopedic M. V. Lomonosov, fated to play an enormous role in the Academy's history. It was Lomonosov, incidentally, who proposed the title of corresponding member for those unable to become full members of the Academy. The Orenburg Kray geographer and historian P. I. Rychkov was elected the first corresponding member in 1759.

In 1728 the Academy began to publish in Latin its scientific organ *Kommentarii* (Commentaries) (Vol 1-14,

1728-1751), then *Novyye kommentarii* (New Commentaries) (Vol 1-20, 1750-1776), *Akty* (Acts) (Vol 1-6, 1778-1786) and *Novyye akty* (New Acts)(Vol 1-15, 1787-1806). In Russian the Academy periodically published its *Kratkoye opisaniye Kommentariyev* (A Brief Account of the Commentaries), the newspaper *Sankt-Peterburgskiye vedomosti* (Saint Petersburg Gazette) and the popular science journal *Mesyachnyye istoricheskiye, genealogicheskiye i geograficheskiye primechaniya v Vedomostyakh* (Monthly Historical, Genealogical and Geographical Notes in the Gazette). In addition, from 1755 to 1764 the Academy published its *Yezhemesyachnyye sochineniya* (Monthly Works), containing interesting historical and ethnographical data. In late 1728 the Academy issued its first *Kalendar'* (Calendar), subsequently an annual publication. In the 19th Century the Academy's periodicals were the *Uchyonyye zapiski* (Learned Notes) and *Izvestiya* (News).

As early as the second quarter of the 18th Century the Russian Academy of Sciences was attracting the attention of scientists in Europe with its mathematical, astronomical and geographical studies. Preeminent among these was cartography, including outstanding contributions by Academicians Joseph de Lisle, Euler and later Lomonosov. From 1727 to 1730 the Academy conducted its first astronomical expedition to establish the geographical position of the northern areas of European Russia. From 1733 to 1743 members of the Academy of Sciences took an active part in Bering's 2nd Kamchatka Expedition and one participant, S. P. Krasheninnikov, compiled his *Opisaniye zemli Kamchatki* (An Account of the Land of Kamchatka). The first teaching atlas and the *Rossiysko-geograficheskiy leksikon* (Russian Geographical Lexicon) were published in 1737. City plans of Petersburg and Moscow were also compiled. The year 1745 saw the publication of the first scientific *Atlas Rossiyskiy* (Russian Atlas) comprising 19 maps of individual localities and one general map. From 1739 all this work was concentrated in the Academy of Sciences' Geographical Department, for a long time Russia's sole geographical establishment.

Until 1747 the Academy of Sciences operated without statutes, since Peter I did not live to compile detailed regulations and the Statutes drafted under Catherine I had not been put into effect. The Academy operated under the statute approved by Peter I, but even this was applied in a curtailed form and the academicians were kept in ignorance of the clause authorizing them to elect their own president. In 1725 the Muscovite physician-in-ordinary L. L. Blyumentrost was appointed the Academy's first president. From 1733 on there were frequent changes of presidents, many of whom resigned to follow a diplomatic career. From 1741 to 1745 the Academy had no president at all and was run personally by Schumacher, a councillor in the Academy's Chancellery who was instrumental in suppressing the above-mentioned clause in the Petrine Statute. In 1745 the 18-year-old Count K. G. Razumovskiy was appointed president; holding this post until 1798.

The Regulations (Statutes) of the Petersburg Academy of Sciences were approved on 24 March 1747. The Humanities Department was detached from the Academy and transferred to the University, leaving the Academy only the natural sciences. Members of the Academy were termed academicians, while those who worked at the University were termed professors, endowed with fewer rights and privileges than the academicians. The members of the Academy were dissatisfied with the new Statutes. Lomonosov, elected to the Academy in 1742, urged its revision and drafted his own "Regulations" and "Privileges" of the Academy of Sciences, but his draft was not approved. Lomonosov was at great pains to develop the Academy's university and high school, of which he was for a while rector. After Lomonosov's death in 1765 the university gradually declined, leaving no record of its demise. At the same time Moscow University, founded in 1755 according to Lomonosov's plan, developed apace.

The 1760's and 1770's were a heyday for mathematical sciences at the Academy, thanks in no small measure to the mathematician Leonard Euler, who at the age of 20 joined the Academy of Sciences shortly after its inception. His works, published in Russia, proclaimed him a great mathematician but he was forced to leave Russia during the infamous sway of the Empress Anna's favorite Biron (1730-40). Euler did not return to Russia until 1766, at the height of his fame. During his absence he maintained his connections with the Academy and printed his works in its *Commentaries*. After his return to Russia, Euler wrote many works on mathematics, geometrical optics and physics. He founded the Russian physics and mathematics school which provided the basis for the development of mathematical training in Russia and which bore fruit in many fields of mathematics long after his death in 1783.

In the 1760's the Academy of Sciences organized extensive geographical and astronomical expeditions to study European and Asian Russia. The plan for these expeditions was drawn up by Lomonosov and carried out in 1768-1774 by P. S. Pallas and S. G. Gmelin, who had come to work in Russia, and the young Russian scientists S.Ya. Rumovskiy, I.I. Lepekhin, V.F. Zuyev and others. The results of these expeditions provided the basis for contemporary knowledge of Russia.

The Academy's translation work, which began shortly after its inception, helped to shape the Russian literary language and Russian scientific and technical terminology. In the 18th Century the Academy of Sciences was the sole center engaged in translating, publishing and disseminating West European literature and scientific works. Russia's first book-shop was established at the Academy.

Extensive changes were made in the Academy's structure in the 1760's and 1770's. In place of the ever-absent president, Count Razumovskiy, a director was appointed. The first director was V.G. Orlov. In 1766 his office, which had administered the Academy, was replaced by a Commission which included academicians. In 1783 Princess Ye. R. Dashkova was appointed director of this commission and in the same year was made head of the newly-founded Russian Academy, which was quite separate from the Academy of Sciences and was charged with "perfecting and enriching the Russian language" and compiling an "explanatory Russian dictionary." Under Princess Dashkova's stewardship the Academy of Sciences' main building in Petersburg was completed to the design of the architect Kvarenga.

In the 1780's and particularly in the early 1790's the Academy of Sciences was affected by Catherine II's domestic and foreign policies under the impact of the French Revolution. The French mathematician Condors, elected an honorary member of the Academy in 1776, was excluded from honorary membership by a decree of Catherine II for his part in France's revolutionary developments, while Dashkova was forced to take indefinite leave for sanctioning the publication of Ya. B. Knyazhin's tragedy *Vadim Novgorodskiy*. Under Paul I Dashkova was induced to sever all connection with the Academy of Sciences, and her place was taken by P. P. Bakunin, himself removed from administration of the Academy in 1798. Instead of the nominal president, Razumovskiy, Paul I appointed Baron A. L. Nikolai, private secretary to the Empress, as new president of the Academy of Sciences. Nikolai set about drafting new statutes but was dismissed in 1803 before they could go into effect. Although the Academy of Sciences had by this time established contacts with many foreign academic institutions and individual scholars and scientists, its work was really in decline, particularly in the field of mathematics and natural sciences.

In the early 19th Century the Academy of Sciences ceased to be Russia's only center of scientific learning. The Tartu (Derpt) University was opened in 1802, Vilnius University (which had existed earlier and which was again abolished in 1832) in 1803, Kazan' University in 1804, Khar'kov University in 1805 and Petersburg University in 1819. In addition, many learned and scientific societies were formed: the Society of Natural Researchers, the Society of Mathematicians and the Society of Russian History and Antiquities - at Moscow University; the Mineralogical Society in Petersburg. These universities and societies became independent scientific centers.

On 25 July 1803 the Academy of Sciences received new Statutes and a new organization. It was finally relieved of its training functions and its "arts" side, but was handed back the humanitarian disciplines (history, statistics, economics and Oriental studies) expropriated under the 1747 Statute. The new Statute defined the Academy of Sciences as the Empire's prime learned association. Its main tasks were to perfect the sciences, adapt "theory and useful research, experiments and observations for practical use" and turn "its works to the benefit of Russia." In addition to its usual basic scientific organ the Academy was also bound "to publish annually in Russian" one volume of notes entitled *Tekhnologicheskiy zhurnal* (Technological Journal). The number of academicians-in-ordinary was increased to 18, and the number of associate academicians to 20. For the first time in the Academy's history, it was entitled to elect new members by ballot in Conference, subject to subsequent confirmation by the Emperor. Academicians were also authorized to participate directly in the administration of the Academy.

However, due to the depreciation of money as a result of the European economic blockade and the Napoleonic Wars, the Academy of Sciences found itself in bad financial straits, which had its effect on both new personnel and on the state of the academy's various institutions, many of which went into a decline. The Physics Center, for example, was in a very poor state, while the Anatomy Theater had to close down. In 1815 the Botanical Gardens were also closed, and the Kunstkammer comprehensive museum was turned into a storehouse for lack of accomodation. Furthermore, from 1810 to 1818 the Academy of Sciences lacked a president and was virtually run by its permanent secretary, Nicholas Fuss.

In January 1818 Count S. S. Uvarov was appointed president of the Academy of Sciences and remained in this post until 1855. From 1833 to 1849 Uvarov was also Minister of Public Education and has gone down in history as the ideologist of what is called "official folkism," preaching the unity of "Orthodoxy, autocracy and folkism." He profited by the centenary of the Academy of Sciences to drum up financial support. Later Uvarov also used his position as Minister of Public Education for the same ends. The Academy's improved finances made it possible to repair existing premises, build extensions and increase staff. It attracted such prominent newcomers as the physicists V. V. Petrov, Emil-Christian Lentz and Moritz-Hermann Jacoby; the mathematicians V. Ya. Bunyakovskiy, M.V. Ostrogradskiy and P. L. Chebyshev; the biologist and anthropologist Karl von Baer; the astronomer V.Ya. Struve, the founder of Pulkovo Observatory; the historians and philologists N. G. Ustryalov, M. P. Pogodin, A. Kh. Vostokov, P. M. Stroyev, I. I. Sreznevskiy, and others. At the same time there arose independent academic museums, offshoots of the former Kunstkammer, which became the nuclei of future research institutes. In 1818, for instance, there emerged the Asian Museum, founded by Academician Christian von Fraehn, which laid the basis for Russian Oriental studies. Then in the 1830's there emerged the Anatomical, Anthropological, Botanical, Zoological, Mineralogical and Ethnographical Museums.

Through the efforts of Academicians V. V. Petrov and Georg-Friedrich Parrot the Physics Center was expanded and provided facilities for Academicians Lentz and Jacoby's researches on electromagnetism and galvanoplastics. Finally, the now prestigious Pulkovo Observatory was founded in 1839 by the Russian astronomer V.Ya. Struve.

In 1836 Uvarov arranged for the adoption of new Academy Statutes, subsequently amended in 1841. The Academy was divided into three divisions: 1) Physics and Mathematics; 2) Russian Language and Literature; 3) History and Philology. The second of these divisions originated from the previously mentioned Russian Academy for the Study of the Russian Language and Literature. The Academy retained this structure for some 90 years, until 1927. President Uvarov favored the idea of self-sufficing science, independent of the country's needs, and this idea was reflected in the new Statutes, which stipulated that the Academy retain exclusively scientific functions. The Academy itself was defined as the "predominant Scientific estate of the Russian Empire."

The question of reforming the Academy of Sciences was again raised in the 1850's and 1860's, and the drafting of new academic statutes lasted from 1857 to 1865. This time discussion of the reforms extended far beyond the Academy of Sciences, being debated among university professors, in the press and in society. The Academy frequently came under fire for its exclusiveness and its tendency to serve "pure science." In 1864 the renowned explorer Count F. P. Litke was appointed president of the Academy, but a more influential functionary at this time was its permanent secretary, the historian and economist K. S. Veselovskiy (1857-1890). Veselovskiy, ever a staunch advocate of pure science, proposed in one of the last drafts that the 1865 Academic Statutes specify that it was the Academy's duty to enrich and perfect science, whereas it was up to society (not the Academy) to put scientific achievements into practice. He failed, however, to get this draft accepted.

By the late 19th Century the Academy of Sciences had actually become a closed scientific establishment. The government unceremoniously interfered in its affairs, encouraged the election of foreigners and did everything in its power to exclude more liberally minded scholars and scientists from the Academy. For example, Moscow University's Professor D. I. Mendeleyev, a corresponding member of the Academy of Sciences from 1876, was blackballed in the 1880 elections of full members. I.M. Semyonov, K.A. Timiryazev and A.G. Stoletov also failed to gain full membership at that time. A group of academicians headed by Litke and Veselovskiy opposed their election, arguing their over-liberal views. Here, of course, they had the government's full backing. Mendeleyev's exclusion from full membership caused a wave of indignation in enlightened society, leading to Litke's replacement as president of the Academy. In April 1882 Count D. A. Tolstoy was appointed to the presidency, combining this with the post of Minister of the Interior and Chief of Police.

The late 19th Century witnessed the rapid development of chemical sciences at the Academy, aided by the outstanding work of Academicians N. N. Zinin and A.M. Butlerov. Zinin was responsible for the discovery of aniline dyes and was co-founder of the Russian Chemical Society. Butlerov, elected academician in 1870, had developed the theory of the chemical structure of organic compounds.

In mathematics significant results in analysis of the theory of numbers, probability theory and the theory of the approximation of functions were achieved by Academician P.L. Chebyshev.

In physics attention was largely concentrated on meteorology. The development of the Russian meteorological service was due in no small measure to the energy and knowledge of Academician G.I. Vil'de. In 1866 the Academy was given control of the Main Physical Observatory, at which marine and meteorological departments were later established. This was followed by the extension of network of meteorological stations, including the establishment of a magnetic and meteorological station at Pavlovsk. In 1862 the Pulkovo Observatory was detached from the Academy of Sciences, but never severed its scientific ties with the latter. The observatory's director and academician-astronomers were appointed by the Academy of Sciences.

In geology the leading researchers were Academician G. P. Gel'mersen of the Chair of Geognosy and

Academician N.I. Koksharev, a specialist in crystallographic ornithognosy. The latter published the multivolume *Materialy dlya mineralogii Rossii* (Material on the Mineralogy of Russia) which paved the way for subsequent Russian and European works on crystallography. In 1886 the Russian geologist A. P. Karpinskiy came to the Academy of Sciences. Co-founder and from 1885-1903 director of the Geological Committee, Karpinskiy was subsequently (from 1917) president of the Academy of Sciences.

The development of biology is linked with the names of such renowned Russian botanists and biologists as A.S. Famintsyn, A.O. Kovalevskiy, I.I. Mechnikov and K.A. Timiryazev, connected with the Academy in the 1880's and 1890's.

Its leading astronomer was Academician O.V. Struve, who discovered over 500 double stars and from 1862 was director of Pulkovo Observatory.

In the humanities such fields as history, Oriental studies, Greco-Roman Antiquities, linguistics and literary history were ably represented by Academicians N.G. Ustryalov, A.A. Kunik, N.V. Kalachov, N.F. Dubrovin, P.V. Nikitin, B.A. Dorn, V.V. Vel'yaminov-Zernov, V.R. Rozen, V.V. Radlov, V.P. Vasil'yev, I.I. Sreznevskiy, Ya.K. Grot, F.I. Buslayev, P.P. Pekarskiy, A.F. Bychkov, S.M. Solov'yov, M.I. Sukhomlinov, A.N. Veselovskiy and I.V. Yagich.

In this period, too, the Academy of Sciences expanded its expeditionary work and its contacts with international scientific organizations. The Academy became a member of many international scientific organizations, societies and commissions which originated in the late 19th Century, including the: International Astronomical Society (1863); International Organization of Archeologists (1867); International Organization of Meteorologists (1872); International Organization of International Law (1874); International Organization of Orientalists (1876); International Organization of Geologists (1881); International Organization of Ornithologists (1884). Academicians increasingly made foreign study tours, attended international and national scientific congresses and conferences and were elected honorary members of foreign academies of sciences and scientific societies. Academicians Butlerov, Struve, Mechnikov and Karpinskiy, for example, were simultaneously honorary members or corresponding members of many foreign academies of sciences and scientific societies. From 1867 Struve was president of the International Astronomical Society. Karpinskiy was elected vice-president of the 4th International Geological Congress in Zurich and for many years represented Russia at international geological congresses. In 1895 the Paris Academy of Sciences awarded him its Cuvier Prize.

In 1889, upon the death of D. A. Tolstoy, the Tsarist Government appointed Grand Duke K.K. Romanov - a relative of the tsar and a well-known poet under the pen name "K. R." - as the new president of the Academy of Sciences. In March 1890 its permanent secretary Veselovskiy retired, to be replaced by Academician A.A. Shtraukh. After a brief incumbency he was replaced in 1893 by the historian N. F. Dubrovin, himself replaced in 1904 by the Orientalist S. F. Ol'denburg, who held the post for 25 years.

The new administration once again raised the question of revising the Academy's Statutes, but the proposals submitted contained no radical changes and the commission appointed to draft new statutes recommended retention of the 1836 Statutes. Instead of new statutes, new personnel levels were approved and the Academy's budget increased. In 1908 the Academy again began to review its staff situation, and this work was not completed until 1912, when the Academy was authorized 153 staff slots. The budget was set at 1,007,000 rubles, most of this for salaries. A mere 47,000 rubles were earmarked for research. In the period from 1890 to 1917 some 68 persons were elected full members of the Academy - 29 of them to the Department of Physics and Mathematics, 20 to the Department of Russian Language and Literature and 19 to the Department of History and Philology. Among these were such world-renowned scholars and scientists as: the mathematicians A. M. Lyapunov, V. A. Steklov and A. N. Krylov; the astronomers F.A. Bredikhin and A.A. Belopol'skiy; the physicists B.B. Golitsyn and M.A. Rykachyov; the chemists V.I. Vernadskiy and N.S. Kurnakov; the geologists F.N. Chernyshyov, A.P. Pavlov and Ye.S. Fyodorov; the botanists S.I. Korzhinskiy, M.S. Voronin and I. P. Borodin; the zoologists D. N. Anuchin, V. V. Zelenskiy and N.V. Nasonov; the physiologist I.P. Pavlov; the historians and philologists V. G. Vasil'yevskiy, A. A. Shakhmatov, A. N. Pypin, F. F. Fortunatov, N. P. Kondakov, A. S. Lappo-Danilevskiy, S. F. Ol'denburg, V.O. Klyuchevskiy, A.I. Sobolevskiy, F.I. Uspenskiy, P.K. Kokovtsov, N.Ya. Marr, V.V. Bartol'd, M.M. Kovalevskiy and N. K. Nikol'skiy. The list shows that by the early 20th Century the Academy of Sciences concentrated almost the entire elite of Russian science.

The scientific personnel of the Academy's various establishments also underwent marked changes in this period and consisted largely of young scientists, often the pupils of established academicians, who lectured at universities and were members of learned societies.

Mathematics in this period was chiefly represented by Academicians Chebyshev and Lyapunov - the latter the founder of the modern theory of stable equilibrium. In addition to basic mathematical research the Academy's mathematicians, and particularly A. A. Markov, did a great deal to disseminate classic mathematical works in Russia. A. N. Krylov translated and published Newton's treatise *Mathematical Principles of Natural Philosophy*.

In 1889 F. A. Bredikhin, regarded as the founder of Russian astronomy, was appointed director of Pulkovo Observatory. Bredikhin developed the mechanical theory of comet forms and the mathematical theory of the origin of meteorite streams. Although he directed the Pulkovo Observatory for only a short while (until 1895), he wrought some far-reaching improvements. The observatory's staff changed, benefiting from the services of such young scientists as A. N. Sokolov, its vice-director, and A. A. Belopol'skiy, senior astronomer for astrophysics. The observatory's equipment and instruments were also greatly improved. In 1900, when O. A. Baklund was director, Bredikhin arranged for the establishment of a branch of the Pulkovo Observatory at Nikolayev. Senior astronomer A.A. Kostinskiy developed the photography of heavenly bodies as a method of observing them. He is rated as the pioneer of astrophotography in Russia and one of the pioneers of this technique in the world.

Aerodynamic research was conducted outside the Academy of Sciences, but the Academy supported people engaged in these studies: K. E. Tsiolkovskiy was given some financial aid; N. Ye. Zhukovskiy was elected a corresponding member, and in the Soviet period S. A. Chaplygin was made a full member of the Academy of Sciences.

B.B. Golitsyn managed to get the Physics Center's budget increased from 1,000 to 2,000 rubles, but even this did not enable the center to be equipped with the then modern equipment. In 1900 a Permanent Central Seismic Commission was established, and the Center received additional premises in the cellar of the Academy's main building. The commission's task was to set up a network of seismological stations. Golitsyn also arranged for the establishment of a system of 1st- and 2nd-category stations and developed a new type of seismograph and a seismometric system. He was a pioneer of the new branch of geophysics - seismology.

In geology important work was being done by Karpinskiy and F. N. Chernyshyov, engaged in a systematic geological survey of Russia. They also founded a national geological museum for scientific research and to popularize geology. In 1900 Chernyshyov was appointed director of the reorganized museum. V. I. Vernadskiy and the young A. Ye. Fersman also made their contribution to geological research. Academician Vernadskiy later raised the importance of studying Russia's deposits of radioactive materials.

In botany S.I. Korzhinskiy, who joined tha Academy in 1893 as an assistant, concentrated on the study of Russian flora and arranged floristic and geobotanical studies in Russia, utilizing the Botanical Museum. The results of his research were summarized in *Gerbariy russkoy flory* (Herbarium of Russian Flora). After Korzhinskiy's death his work was continued by I.P. Borodin (an academician from 1902), who published his *Flora Sibiri i Dal'nego Vostoka* (The Flora of Siberia and the Far East). M.S. Voronin, who worked for a time at the Botanical Museum, laid the basis of botanical mycology. Borodin and Voronin used their own money to found in Bologoye Russia's first biological station to study the vegetation of fresh-water reservoirs. The station was later transferred to Lake Seliger, and hence to Karelia.

The Laboratory of Plant Anatomy and Physiology,

directed by Academician A.S. Famintsyn, was founded at the Academy in 1890. In 1907 I.P. Pavlov was appointed director of the laboratory, which then concentrated on the study of conditioned reflexes, making Pavlov's fame. From 1891 Pavlov also headed the Physiological Department of the Institute of Experimental Medicine.

The Zoological Museum, well endowed in the late 19th Century, was reorganized in 1901 and was defined in its statutes as "the central institution for knowledge of the animal kingdom, primarily in Russia." By the late 19th Century the morphological trend in zoology was well established and a special Zoological Laboratory was set up at the Zoological Museum.

Byzantine studies at the Academy were represented by Academicians V. G. Vasilevskiy, N. P. Kondakov and F. I. Uspenskiy. In 1894 the Russian Archeological Institute was founded in Constantinople to direct Russian research into the history and ancient monuments of Greece and the Near East. The institute was directed by Academician Uspenskiy.

Oriental studies proceeded along the lines mapped out by V. R. Rozen, who from 1885 headed the Oriental Department of the Russian Archeological Society. The Rozen school furnished such outstanding Orientalists as Academicians V.V. Bartol'd, S.F. Ol'denburg, N.Ya. Marr and I.Yu. Krachkovskiy. Russia's eminent place in Oriental studies is demonstrated by the fact that in 1899 the 12th Congress of Orientalists in Rome decided to found an International Union for the Study of Central and Eastern Asia, with its central committee in Petersburg.

In the field of Russian language and literature research centered on Slavic Russian philology, as expounded in the works of Academicians A. A. Shakhmatov, F. F. Fortunatov, A. I. Sobolevskiy, I. V. Yagich and V.I. Lamanskiy. One of the main tasks was the compilation of the *Slovar' russkogo yazyka* (A Dictionary of the Russian Language). A study of extant dialects was also begun and in 1903, on Shakhmatov's initiative, the Department of Russian Language and Literature incorporated the Moscow Dialectological Commission. From 1903 to 1917 this commission collected extensive data on dialects and compiled dialectical maps of Russian and other Slavic languages on Russian territory. From

1908, under the editorship of I. V. Yagich, works on Slavic studies were printed in the *Entsiklopediya slavyanskoy filologii* (Encyclopedia of Slavic Philology). From 1875 Shakhmatov also published the best Slavic journal, entitled *Archiv fuer slavische Philologie*. In addition to his research on Russian chronicles and the history of Russian, Academician Shakhmatov was also a staunch champion of linguistic freedom. Under the difficult conditions of the period he defended freedom of speech and the national languages' right to recognition and use in Russia.

The idea of establishing an institution for studying the life and work of Pushkin originated in 1899 and culminated in 1905 with the founding of the Pushkin Museum and Pushkin Center, a project which was finally completed in 1918. Academicians N. A. Kotlyarevskiy and B. L. Modzelevskiy were the founders and organizers of the Pushkin Center, which from 1903 began to publish a special organ entitled *Pushkin i yego sovremenniki* (Pushkin and His Contemporaries).

In the late 19th and early 20th Centuries the Academy organized several expeditions to study the lands of European Russia and of Siberia along the coast of the Arctic Ocean. The first of these was the geologist I. D. Cherskiy's expedition, continued after his death by E. V. Tol', curator of the Geological Museum. Then came Tol's own expedition, followed by an undertaking that has come to be known as the Russian Polar Expedition. Finally the 1899-1901 major expedition to survey Spitsbergen, which continued V. Ya. Struve's work of the 1830's in measuring the arc of the meridian. From 1903 to 1913 Karpinskiy and Chernyshyov worked on the publication of an international geological map.

The Academy of Sciences' publishing work was expanded and diversified with serial and periodical works linked to special disciplines. The *Trudy Botanicheskogo muzeya* (Works of the Botanical Museum), *Trudy Botanicheskoy laboratorii* (Works of the Botanical Laboratory), *Yezhegodnik Zoologicheskogo muzeya* (Yearbook of the Zoological Museum), *Vizantiyskiy vremennik* (Byzantine Times) and other similar periodicals began to appear separately. In 1901 the Bureau of International Bibliography was founded to publish the *Russkaya bibliografiya po yestestvoznaniyu i matematike* (Russian

Bibliography on the Natural Sciences and Mathematics).

In the early years of the 20th Century the Academy of Sciences intensively developed its international scientific connections. Its members played an active part in international congresses of geologists, geographers, historians, Orientalists, archeologists and Americanists and functioned on the International Aeronautical Commission, the International Seismological Association and many other learned bodies.

World War I had an adverse effect on the Academy's development. Funds were reduced and several of the Academy's institutions, including the Sebastopol Biological Station, had to suspend work. The Academy was forced to tailor its work to war needs. Academician Golitsyn, for example, organized at the Main Physical Observatory a workshop for the production of devices and instruments for the army. The workshop even produced clinical thermometers. On 21 January 1915 a group of academicians led by V. I. Vernadskiy proposed that the Academy establish a permanent commission for the study of Russia's natural production resources. The commission was founded on 4 February 1915 and consisted of Academicians A. S. Famintsyn, A.P. Karpinskiy, B.B. Golitsyn, M.A. Rykachyov, V.V. Zelenskiy, I.P. Borodin, V.I. Vernadskiy, N.V. Nasonov, I. P. Pavlov, N. S. Kurnakov, I. I. Andrusov and V. I. Palladin. Other leading Soviet scientists, particularly D.N. Anuchin, V.A. Obruchev and Ye.S. Fyodorov, were subsequently coopted to the Commission. The commission's function was to publish a work describing Russia's known natural resources and make special studies of individual little-known natural deposits for the use of industry. The commission received funds for the publication of the compendium *Yestestvennyye proizvoditel'nyye sily Rossii* and for individual studies. Together with the commission's regular budget, these funds amounted to 197,700 rubles.

In the last decade before the 1917 October Revolution the Academy of Sciences made major contributions to the development of natural sciences in Russia.

2. The Soviet Period

In May 1917, following the February Revolution, the Provisional Government renamed the Imperial Academy the Russian Academy of Sciences. At its General Assembly on 15 May 1917 the academicians for the first time in the Academy's history elected their own president - the geologist A. P. Karpinskiy. The botanist I. P. Borodin was elected vice-president, while the Orientalist S. F. Ol'denburg retained his post as permanent secretary.

After seizing power in October 1917 the Bolsheviks were naturally anxious to involve cultural figures, scientists and engineers in the formation of the Soviet state system. Here, of course, the Academy of Sciences played a leading role. The Bolshevik leaders, and particularly Lenin, realized that in this sphere there could be no such abrupt break as there had been in the country's political and economic life. On the one hand, they wanted to "take all science, and technology, all knowledge and art" but they realized, Lenin claimed, that this science, technology and art were "in the hands of specialists and in their heads." Furthermore, these specialists were either members of the various non-Communist parties or else were guided by the idea of "pure science" and had no interest in political and social affairs.

In view of all this, Lenin insisted that the leading functionaries of the People's Commissariat of Education, and particularly Lunacharskiy, handle scientists and scholars and especially the Academy of Sciences very carefully. His directive read: "Treat the Academy scrupulously and carefully and only gradually, without injuring its organs, introduce it more firmly and organically into the new building of Communism." When Lenin learned that the People's Commissariat of Education was preparing a plan for "bold" reorganization of the Academy of Sciences, he immediately ordered it to desist from such an experiment.*

---

* A. V. Lunacharskiy, "On the 200th Jubilee of the All-Union Academy of Sciences," *Novyy mir*, 1925, No 10, p 110.

The People's Commissariat of Education initiated official contacts with the Academy of Sciences in January 1918. In April 1918 N. P. Gorbunov, secretary of the Council of People's Commissars, was also commissioned by Lenin to establish direct contacts with the Academy. The idea was to get the Academy of Sciences to participate in Soviet construction by drafting extensive plans for scientific research and practical projects utilizing the country's resources to boost the economy. The Academy's executive, led by President Karpinskiy, was not averse in principle to conducting useful work for the welfare of Russia. On the contrary, it agreed to "tackle the feasible scientific and theoretical solution of individual tasks posed by the needs of state construction" and in connection with this instructed the Commission for the Study of Russia's Natural Production Resources to draft appropriate notes for a program of research and practical work in the field of minerals, power engineering, agriculture and so forth. At the same time it made it quite clear that the Academy of Sciences would continue to adhere to its scientific traditions, refused to work to the government's dictates and declined to compile unrealistic extensive plans for research and practical work. In passing, it drew attention to the "gulf" which had developed as a result of the Bolsheviks' October Coup and which prevented "the development of real continuity, which alone can constitute a reliable pledge of vital creativity."*

On 18 April 1918 the Council of People's Commissars discussed these proposals of the Academy of Sciences and decided to "accomodate" them and "recognize in principle the need for financing appropriate works of the Academy of Sciences and point out to it, as a particularly important and urgent task, a solution to the problems of the correct distribution of industry in the country and the most efficient utilization of its economic resources."** It was then that Lenin drafted his "Outline Plan of Scientific and Technical Work" which proposed, among other things, that the

---

* G.A. Knyazev and A.V. Kol'tsov, *Kratkiy ocherk istorii Akademii nauk SSSR* (A Brief Outline History of the USSR Academy of Sciences), 1964, pp 71-74.
** I.S. Smirnov, *Lenin i sovetskaya kul'tura* (Lenin and Soviet Culture), 1960, p 262.

Supreme Sovnarkhoz instruct the Academy of Sciences "to form a number of commissions of specialists to compile as soon as possible a plan for the reorganization of industry and the economic development of Russia."*

This marked the start of economic cooperation between the Russian Academy of Sciences and the Soviet regime. In the field of social sciences and ideology Lenin's government did not trust the academicians, and in 1918 it founded a separate Socialist Academy (from April 1924 the Communist Academy), which existed until 1936.

From 1918 the Academy of Sciences' material base grew stronger with each passing year. Even during the Civil War the Soviet government frequently assigned it funds and helped it to publish scientific works. In 1924 and 1925 the USSR Council of People's Commissars assigned the Academy 500,000 gold rubles in addition to its regular budget. Its scientific facilities also improved. During the Civil War the Academy of Sciences established new commissions, while the Commission for the Study of Russia's Natural Production Resources founded 20 branch departments. The Institute of Physicochemical Analysis and the Institute for the Study of Platinum and Other Precious Metals were founded in 1918. In the 1920's there appeared the Physics and Mathematics Institute and the Radium, Seismological, Physiological and Soil Institutes, as well as laboratories of biochemistry and biogeochemistry. The Pushkin Center (Institute of Russian Literature), Tolstoy Museum, Archeographic Commission and other establishments were also transferred to the Academy. All in all, whereas prior to 1917 the Academy had five laboratories, five museums, one institute (the Caucasian History and Archeological Institute), two observatories and 15 commissions, by 1928 it had nine institutes, three laboratories, seven museums and 20 commissions. In the period from 1917 to 1925 some 29 new members were elected, but by 1925 the total number of full members was restricted to 42. The Academy's scientific staff increased from 154 in 1917 to 413 in 1925. Including technical personnel some 1,055 persons

---

* V.I. Lenin, *Polnoye sobraniye sochineniy* (Complete Collected Works), 5th ed, vol 36, 1962, pp 228-231.

worked at the Academy of Sciences in 1925.

In 1919 the Academy of Sciences began its study of the Kursk Magnetic Anomaly under the supervision of Academician P.P. Lazarev. Studies of the Kola Peninsula began in 1920 and led to the establishment of an industrial center to work the local apatite-nepheline deposits. Between 1919 and 1925 the Academy organized over 150 scientific expeditions to areas of the Urals, Siberia, Central Asia, the Caucasus and Northern and Southern European USSR. Geological, mineralogical, geophysical and hydrological research was conducted, as well as studies of the soils, the flora and the fauna. A number of linguistic, anthropological and statistical economic expeditions as well as four comprehensive expeditions were also conducted. These were directed by the special Commission for Scientific Expeditions, established in 1921.

Major studies completed by Academicians in this period include: A.Ye. Fersman's *Geokhimiya Rossii* (The Geochemistry of Russia) (1922), V. A. Steklov's *Osnovnyye zadachi matematicheskoy fiziki* (The Basic Tasks of Mathematical Science) (1922 - 23), I. P. Pavlov's *Dvadtsatiletniy opyt ob'yektivnogo izucheniya vysshey nervnoy deyatel'nosti (povedeniya) zhivotnykh* (Twenty Years' Experience in the Objective Study of the Higher Nervous Activity (Behavior) of Animals) (1923), V. I. Vernadskiy's *Istoriya mineralov zemnoy kory* (The History of the Minerals of the Earth's Crust) (Vol 1, 1925) and Г. Yu. Levinson-Lesing's *Petrografiya* (Petrography).

In September 1925 the Academy of Sciences celebrated its 200th anniversary with ceremonies attended by some 1,000 delegates from the USSR and from universities and academies in Western Europe, America, Asia and Australia. The Soviet government was represented by M. I. Kalinin, chairman of the Central Executive Committee.

On 27 July 1925, just prior to this jubilee, the USSR Central Executive Committee and Council of People's Commissars adopted a resolution which recognized the Russian Academy of Sciences as "the highest learned establishment of the USSR." In other words, the Soviet government recognized the Academy as the center of Soviet science. Instead of a Russian it was made an

All-Union body under the title Academy of Sciences of the Union of Soviet Socialist Republics - the USSR Academy of Sciences, for short.

Later, on 18 June 1927, the USSR Council of People's Commissars approved new Academic Statutes. The Academy was now divided into two divisions - the Division of Mathematical and Natural Sciences and the Division of Social Sciences. The Academy's tasks were: a) to develop and perfect the scientific disciplines over which it has competence, enriching them with new discoveries and research methods; b) to study the country's natural production resources and promote their use; c) to adapt scientific theories and the results of scientific experiments and observations for practical use in industry and the cultural and economic construction of the USSR.* Under the new Statutes the number of full members of the Academy was to be raised to 75 and their election organized as follows: candidates were to be nominated by scientific establishments, groups of scientists and social organizations; elections were to be preceded by reports on the candidates submitted by a special commission of the country's academicians and scientists.

This paved the way for a crucial reorganization of the Academy of Sciences and broke its tradition as a closed and more or less autonomous corporation of scholars and scientists through the sovietization of its membership and staff. Hitherto the Academy had been "slowly coming to face socialist construction" but this tardiness proved intolerable to the Soviet authorities in their efforts to build a material basis for socialism. Radical reorganization of the Academy of Sciences began in 1929 and 1930.

In 1929 the USSR Council of People's Commissars sent a special commission "to aid the Academy of Sciences." The commission was instructed to scrutinize the academy's staff and inspect the work of the academy's institutions. This led to the dismissal of several staff members "who did not conform to their ap-

---

* *Sobraniye zakonov i rasporyazheniy Raboche-krest'-yanskogo pravitel'stva SSSR* (Collected Laws and Regulations of the USSR Workers and Peasants' Government), 1927, No 35, p 367.

pointment" and even to the arrest and exile of several academicians, such as the historians S. F. Platonov, M.K. Lyubavskiy, Ye.V. Tarle and N.P. Likhachyov. Before the year 1929 was out extensive elections of full and corresponding members were held with "public" participation. Forty-two academicians were elected, including many Marxists such as the power engineer G. M. Krzhizhanovskiy, the geologist I.M. Gubkin, the philosopher A.M. Deborin and the historians M.N. Pokrovskiy and N. M. Lukin. Similar large-scale elections were repeated in 1932, when "persons directly involved in the building of socialism" were admitted to the Academy. Prior to this, in 1930 and 1931, a few individuals had been elected academicians.

A new Presidium of the Academy of Sciences was elected in 1930. Karpinskiy remained president, Academicians V. L. Komarov and G. M. Krzhizhanovskiy were elected vice-presidents and the Marxist historian V.P. Volgin was elected permanent secretary. S. F. Ol'denburg, the previous permanent secretary, was made director of the newly founded Institute of Oriental Studies. On 23 May 1930 the Presidium of the USSR Central Executive Committee also approved the new Academic Statutes. The USSR Academy of Sciences was now commissioned to organize the country's entire research work and its basic tasks were defined as follows: "The Academy of Sciences operates in all fields of theoretical knowledge, roundly promotes the development of research thinking, unites all the basic disciplines, sponsors the development of a single scientific method based on a materialistic world outlook, systematically guiding the entire system of scientific knowledge toward satisfaction of the country's socialist reconstruction needs and the further growth of the socialist social system."

In addition to absorbing these reorganizational measures, the Academy of Sciences in 1930 embarked on the planning of scientific work and research, compiling its first research plan (for 1931) and then a research plan for the period of the 2nd five-year plan (1933-37). Many prominent scientists opposed this practice on the grounds that the specified tasks and planning tied the hands of scientific establishments and fettered individual scientific initiative. However, the government directives prevailed, as did the Academy's burgeoning Party organization which campaigned lustily for the adoption of the research plans.

The structural reorganization of the Academy, which had begun in 1929, eventually led to the formation of individual groups or teams for allied disciplines and theoretical problems and to the abolition or amalgamation of many establishments, which were then converted into institutes. For example, physicomathematical, chemical, geological, technical and biological groups emerged in the Division of Physico-Mathematical and Natural Sciences, and in the Division of Social Sciences - groups on language and literature, sociology, history and economics and Oriental studies. In subsequent years these groups increased in number. In 1930 the Institute of Oriental Studies was formed from the Asian Museum, the Collegium of Orientalists, the Institute of Buddhist Culture and the Turkological Center, while the Commission for the Study of Natural Production Resources, the Commission for Expeditionary Research and the Commission for the Study of the Union and Autonomous Republics together provided the basis for the new Council for the Study of Production Resources. Next year, 1931, saw the foundation of the Botanical Institute (from the Botanical Gardens and the Botanical Museum), the Zoological Institute (from the Zoological Museum), the History and Archeographical Institute (from the History and Archeographical Commission) and the Power Engineering Institute (from the Department of Power Engineering, Commission for the Study of Natural Production Resources). In 1933 and 1934 this was followed by the establishment of the Institute of Genetics, the Steklov Mathematics Institute, the Lebedev Physics Institute, the Institute of Physical Problems, the Institute of General and Inorganic Chemistry, the Institute of Organic Chemistry, the Institute of Combustible Minerals and the Institute of Plant Physiology.

Altogether in 1934 the USSR Academy of Sciences had 25 institutes, 10 laboratories, 11 commissions, one museum and a number of experimental stations. Moreover, in 1932 there had been established the Ural, Far Eastern and Transcaucasian Branches of the Academy, as well as the Kazakhstan and Tadzhik Research Bases. The Kola Research Base was established in 1934. In 1934 the USSR Academy of Sciences had a budget of 25 million rubles, as against 3 million rubles in 1928.

On 14 December 1933 a resolution of the USSR Central Executive Committee transferred the USSR Academy of Sciences to the jurisdiction of the USSR Council of People's Commissars. The aim of this was to "further link the Academy of Sciences' work with the practice of socialist construction and to establish systematic and close cooperation between it and the people's commissariats and the State Planning Commission. To this end the USSR Council of People's Commissars decided on 25 April 1934 to transfer the Academy of Sciences from Leningrad to Moscow. In 1934, as a result of this decision, the Academy's Presidium and most of its main institutions found themselves in Moscow. In November 1935 the new Statutes were adopted.

These Statutes defined the Academy's tasks still more specifically. Its basic task was the systematic use of scientific achievements to promote the building of socialism. Its efforts should be concentrated on the major, outstanding problems of science in all its fields, on studying the country's natural wealth and production resources and on scientific and economic achievements in general to promote their timely and efficient application. In addition, the Academy was to train qualified scientific personnel for the country's national economy and to serve the governmental organs with scientific expertise. The Academy's former technical group was taken as the basis for a new department - the Department of Technical Sciences. The Statutes specified that: "The USSR Academy of Sciences consists of full members (academicians), honorary members, corresponding members and learned personnel working in the Academy of Sciences' institutions." Full members were elected from the ranks of researchers who had enriched science with works of prime importance; corresponding members were elected from major specialists in the various branches of knowledge; honorary members were (Soviet and foreign) scholars and scientists who had made discoveries of world significance. As for the system of electing full members, the new Statutes entrusted preliminary discussion of the candidates' suitability to appropriate academic branch groups. Naturally, these rules were not always fulfilled; the academicians were often by-passed to gain membership for some Party nominee.

In February 1936 the USSR Council of People's Commissars and Central Committee of the All-Union Communist Party (Bolsheviks) decided on the abolition of the Communist Academy and the transfer of its institutions - the institutes of history, philosophy, law, economics, world economics and world politics, as well as the Library - to the USSR Academy of Sciences. The Academy was thus further expanded by the incorporation of these institutions. The new Institute of History, incidentally, was an amalgam of the Communist Academy's Institute of History and the Academy's former History and Archeographical Institute. Its membership was also increased by the addition of Marxist scholars who had worked at the Communist Academy. In 1937 the Academy of Sciences also gained the State Academy for the History of Material Culture, which was turned into the Institute for the History of Material Culture (from 1959 the Institute of Archeology), and in 1938 it acquired the Gorky Institute of World Literature (founded in 1932) and the Research Institute for the Language and Literature of the Peoples of the USSR. In addition, in the period from 1935 to 1939 new institutes were set up at the Academy of Sciences: the Institute of Microbiology in 1935; the Institute of Biochemistry in 1935; the Institute of Evolutionary Morphology and the Institute of Physiology in 1936; the Institute of Theoretical Geophysics and the Institute of Geological Sciences in 1937; the Institute of Mining, the Institute of Machine Science, the Institute of Metallurgy and the Institute of Mechanics in 1938; the Institute of Automation and Telemechanics and the Institute of Permafrost Studies in 1939.

The USSR Academy of Sciences' branches and research bases were also expanded. In 1935 the Transcaucasian Branch was broken up into three new branches - the Azerbaidzhani, Armenian and Georgian Branches. The Uzbek Branch was founded in 1939 and the Turkmen Branch in 1940. The former Kazakhstani and Tadzhik research bases were elevated to branches, and in 1936 the Northern Research Base was established. By 1941 the Academy of Sciences had seven branches (Azerbaidzhani, Armenian, Kazakhstani, Tadzhik, Turkmen, Uzbek and Ural) and two bases (Kola and Northern). By this time the Georgian Branch had been turned into the Georgian Academy of Sciences. In 1941 there was also founded the Lithuanian Academy of Sciences.

In 1938 the USSR Council of People's Commissars amended the structure of the Academy of Sciences, abolishing its branch groups and using them as a basis for new divisions. The Academy now had eight divisions: Physical and Mathematical Sciences, Chemical Sciences, Geological and Geographical Sciences, Biological Sciences, Technical Sciences, Economics and Law, History and Philosophy, and Literature and Language.

In 1935, in order to train top-level scientific personnel, the Academy of Sciences introduced a doctorate system (a postgraduate training system had been introduced much earlier, in 1929). Then, in late 1938, it established another postgraduate system for training doctors and candidates of sciences while the trainees continued at their normal jobs. As a result of this the Academy of Sciences postgraduate trainees increased from 117 persons (81 of them studying for a doctorate) in 1938 to 1,162 persons (236 of them working for a doctorate) in 1941. Those who studied while continuing their jobs tended to aim for candidate of science degrees, while those who studied full-time worked mainly for doctorates.

In May 1935 the USSR Academy of Sciences gained 15 new full members. In October 1939 a further 56 full members and 102 corresponding members were elected. By 1941 the Academy had a total of 120 academicians, 182 corresponding members and a staff of some 8,000, of whom 3,700 were scientific personnel. There were 402 doctors of science, and 1,271 candidates of science. From 1937 to 1939 many of the Academy's scholars and scientists fell victim to the Stalin Terror.

In December 1936, after Karpinskiy's death, the botanist V.L. Komarov was elected the new president of the USSR Academy of Sciences. Academicians I.M. Gubkin and E. V. Britske were elected vice-presidents. From 1935 the technological engineer N.P. Gorbunov was permanent secretary, and the various departments were headed by academic secretaries.

In 1938 the Academy of Sciences had a total budget of 126,700,000 rubles, in 1939 - 158,000,000 and in 1940 176,912,000 rubles.

The 1930's marked the vigorous development of the Academy's expeditionary work, encompassing all fields of research but concentrating on the study of the country's production resources. Its international scientific ties increased until 1937 and then declined due to the aftermath of the infamous Soviet-German Non-Aggression Pact and the raging Stalin Terror. This period was marked by notable achievements in mathematics: L. S. Pontryagin on the theory of groups, M. V. Keldysh on the theory of sets, M.A. Lavrent'yev on the theory of complex variable functions and S. L. Sobolev on the theory of differential equations. In 1937 Academician I. M. Vinogradov published his *Novyy metod v analiticheskoy teorii* (A New Method in Analytical Theory), outlining his method of solving Goldbach theorems for odd numbers. The physicists made some progress on the theory of atomic nucleus structure, the nature of elementary particles, quantum electrodynamics, the physics of crystals and physical optics, photoluminescence, oscillation theory, the physics of dielectrics and molecular physics. In 1934 P.A. Cherenkov derived experimental data on a new type of visible luminescence of pure liquids under the effect of radiation from radioactive substances. These findings were theoretically substantiated by I. Ye. Tamm and I. M. Frank. In 1958 the authors of this discovery were awarded a Nobel Prize. In 1938 Academician P. L. Kapitsa developed a new method of compressing gases. Academician N. N. Semyonov, for his part, laid the basis for the theory of branching chain reactions and the thermal theory of ignition and combustion. For this research he and the British scientist Hinchelwood were awarded a 1956 Nobel Prize. Academician A. Ye. Favorskiy directed research into chemical processes connected with the formation of high-molecular compounds, leading in 1940 to a means of synthesizing vinyl esters from acetylene and alcohols. From 1939, under the supervision of I. V. Kurchatov, the Radium Institute worked on the building of a 100-ton cyclotron, which was only prevented from going into operation by the blockade of Leningrad.

The development of the social sciences was extremely circumscribed in this period. One factor was Stalinist dogmatism which fettered scholars' creative approach and contradicted the basic principles of the materialistic interpretation of historical and cultural processes. Moreover, this dogmatism was aggravated by abnormal forms of Soviet (in effect, Russian) patriotism, which led to idealization, the varnishing of reality and even the blatant falsification of historical truth in relation to various processes and events. The most striking development was the fate which befell Academician M. N. Pokrovskiy's Marxist historical school.

On 23 June 1941, the day after the Germans launched their invasion of the USSR, the Presidium of the USSR Academy of Sciences held an emergency meeting and resolved to review the research program of all the Academy's institutions in order to adapt them to wartime requirements.

Some 2,000 members of the Academy's staff left for the front, and the Academy of Sciences found it very difficult to continue its functions. Many of its institutions were evacuated deep into the Soviet Union and for the first two years were sited at 16 different localities. Conditions in Leningrad were particularly difficult, but the Academy's institutions there continued to operate right through the blockade. On 1 April 1942 Academician I. Yu. Krachkovskiy assumed the chairmanship of the newly-founded Joint Learned Council to administer those institutions of the Academy of Sciences which remained in Leningrad. In late July and early August 1941 the Radium Institute, Physical Engineering Institute and Institute of Chemical Physics, which were working entirely on war needs, were evacuated from Leningrad.

In August and September 1941 the Academy of Sciences compiled its first wartime work program, including 200 projects connected with the supply of material to the army and navy. In September 1941 the Commission to Mobilize the Resources of the Urals was formed from the Academy of Sciences' Ural Comprehensive Expedition. President Komarov himself was chairman of this commission, with Academician I. P. Bardin as his deputy. In early 1942 the commission was renamed the Commission to Mobilize the Resources of the Urals, Western Siberia and Kazakhstan for National Defense. In June 1942 there was founded the Commission to Mobilize the Resources of the Middle Volga and Kama Area. In conjunction with the Institute of Combustible Minerals, the Institute of Geological Sciences and the Bashkir Comprehensive Expedition the commission drafted measures for a considerable expansion of oil production in the No 2 Baku Fields. The Commission for Anti-Tank Devices, the Military Geography Commission, the Aerial Survey and Camouflage Commission, the Geological and Geographical Service Commission, the Marine Hydrophysical Laboratory and the Military Health Commission, founded in 1941-42, also worked on defense projects.

Under the direction of Academician S. I. Vavilov the Academy's physicists designed valuable optical instruments for war needs. The Academy of Sciences' Laboratory for Nuclear Engineering began work in Moscow in 1943 under the direction of Academician I. V. Kurchatov. Under Academician P. L. Kapitsa the staff of the Institute of Physical Problems worked on the mass production of liquid oxygen. The Institute of General and Inorganic Chemistry developed methods of extracting platinum, palladium, rhodium and iridium, organized the production of new heat-resisting and high-capacity alloys at a number of plants and worked out the technology for the production of salt alloys as substitutes for saltpeter. The Academy's specialists also studied agricultural problems, particularly ways of increasing crop yields and supplying the rear areas with food.

During the war the Presidium of the USSR Academy of Sciences was at first located in Kazan' and then, in 1942, was transferred to Sverdlovsk. Throughout May, June and October 1943 Academy institutions were gradually returned to Moscow, the Leningrad institutions returning somewhat later. In Moscow the first session of the reestablished Academy of Sciences was held in late September 1943, summarizing its work in the first years of the war and outlining its work plan for 1944.

During the war the Academy gained several new institutions. In 1942 it acquired the Pacific Institute, followed in 1944 by the founding of the Institute of Art History, the Institute of Russian Language and the Institute of Forestry, then by the addition of the Institute of the History of Natural Sciences in 1945. The Kirghiz and the West Siberian Branches were founded in 1943, and the former Northern Study Base was turned into the Komi Research Base. Respective branches of the USSR Academy of Sciences were expanded in 1943 into the Armenian and Uzbek Academies of Sciences and in 1945 the Azerbaidzhani Academy of Sciences. The Ukrainian and Belorussian Academies of Sciences had emerged independently much earlier, in 1919 and 1929.

By 1945 the USSR Academy of Sciences possessed 53 institutes, 16 laboratories, 35 stations, 31 commissions and 15 museums. On 1 January 1945 it had 142 academicians, 200 corresponding members and a staff of over 4,000. In May 1942 a further five full members, and in September 1943 some 36 academicians and 58 corresponding members were elected.

During the war the USSR Academy of Sciences celebrated several jubilees, including the 250th anniversary of the birth of Voltaire and the 300th anniversary of the birth of Newton. In connection with the latter anniversary the Academy published a special compendium on Newton, while the Royal Society in London donated to it Newton's work *The Mathematical Principles of Natural Philosophy*, published in 1687, and the draft of Newton's letter to Aleksandr Men'shikov, the first Russian to be elected a member of the Royal Society. In June 1945 the Academy celebrated its own 220th anniversary in ceremonies attended by 1,200 persons, including 123 scientists and scholars from 19 foreign countries.

After the war the Academy of Sciences turned all its attention to restoring industry and agriculture. Research on atomic energy was also given priority, leading in 1949 to the testing of the first Soviet atomic bomb, followed by a thermonuclear device in 1953 and the construction in 1954 of the world's first 5,000-kilowatt industrial atomic power station. Computer research culminated in 1952 with the operation of the BESM, the Soviet Union's first electronic computer, developed by a team under Academician S. A. Lebedev.

Despite postwar difficulties and disruption, in 1947 the Academy of Sciences received additional funds for the construction of new buildings, the acquisition of foreign equipment and to increase wages and salaries. The Academy's structure was again expanded by the establishment of new institutions. There emerged, for example, the Kazan', Bashkir and East Siberian Branches and the Daghestani, Yakutsk and Sakhalin Research Bases, which were in their turn reorganized into branches in 1949. Of the newly-founded institutes one might mention: the Institute of Oceanology (1946); the Institute of Petroleum (1947); the Institute of Geochemistry and Analytical Chemistry (1947); the Geophysical Institute (1947); the Institute of Slavic Studies (1947); the Institute of Silicate Chemistry (1948); the Institute of Animal Morphology (1949); the Institute of Higher Nervous Activity (1950); the I.P. Pavlov Institute of Physiology (1950); the Institute of Linguistics, compounded in 1950 from the Institute of Language and Thought and the Institute of Russian Language. In the period from 1951 to 1955 it was further expanded by the addition of 24 scientific establishments. Former branches were turned into the Kazakh Academy of Sciences (1946), the Tadzhik and Turkmen Academies of Sciences (1951) and the Kirghiz Academy of Sciences (1954). The Latvian and Estonian Academies of Sciences were also founded in 1946. In 1947, in connection with this decision of the USSR Council of Ministers, the USSR Academy of Sciences' Presidium formed a special council to coordinate the work of the republican academies of sciences and its own various branches. The council met in annual session to discuss the work plans compiled by the republican academies of sciences and the USSR Academy of Sciences' branches in an attempt to prevent overlapping projects and coordinate general research. It should be noted that in this period the scientific work of the republican academies of sciences was very limited in scope and they continued to be, in effect, mere branches of the USSR Academy of Sciences.

Finally, in 1954, the Division of History and Philosophy and the Division of Economics and Law of the USSR Academy of Sciences were reorganized as the Division of Historical Sciences and the Division of Economic, Philosophical and Legal Sciences.

Back in 1946 the USSR Academy of Sciences had gained 46 new full members and 112 corresponding members. In October 1953 51 academicians and 148 corresponding members were also elected. In 1955 the Academy of Sciences had 151 academicians, and 325 corresponding members; the number of doctors of science had risen to 1,201, and candidates of science to a massive 5,059.

In July 1945 Academician Komarov resigned on medical grounds as president of the USSR Academy of Sciences, his place being taken by S. I. Vavilov. After Vavilov's death in February 1951 Academician A.N. Nesmeyanov was elected president of the Academy.

Like the immediate pre-war period, so the first years after the war were marked by an almost complete cessation of the USSR Academy of Sciences' international scientific ties. These ties were virtually limited to contacts with scientists in the People's Democracies and to speeches and reports by Soviet scientists at various international Communist forums. This was due, of course, to the Cold War and to the official Soviet attitude of disdain for "bourgeois science," which Soviet propaganda termed "pseudo-scientific." It was not until Stalin's death that there was any radical change in this situation, brought about by the new leadership's policy of peaceful coexistence with countries of different socio-economic systems.

Stalinist dogmatism in the field of the social sciences reached its peak in the late 1940's and early 1950's. Preconceived notions based on quotations from the classic works of Marxism-Leninism and the shuffling of facts to meet these ideas formed the bulk of economic, philosophical, juridical and historical studies. Their scientific and academic value was almost nil, and social sciences generally were in a state of decline.

The late 1950's were marked in the Academy of Sciences' work by great achievements in the conquest of space. On 4 October 1957 the USSR launched the world's first artificial earth satellite. Shortly after this, on 14 April and 15 May 1958, two more satellites were launched, one of them bearing the test dog Layka. These flights yielded valuable data on the upper layers of the atmosphere, cosmic radiation and so forth. The launching of these satellites also emphasized the major strides that had been made in rocket engineering.

At the Institute of Atomic Energy a team of physicists under the supervision of the institute's director, Academician I. V. Kurchatov, continued its successful research on controlled thermonuclear reactions. The results of this research into powerful pulse charges in gas media to obtain high-temperature plasma were published in 1956 and 1957 and brought

Academicians L.A. Artsimovich and M.A. Leontovich and their team a 1958 Lenin Prize. Lenin Prizes were also awarded to Academician N. N. Bogolyubov for developing a new method in quantum field theory and statistical physics and to Academician Ye. K. Zavoyskiy for his discovery and study of paramagnetic resonance.

As we have mentioned earlier, in 1956 Academician N. N. Semyonov received a Nobel Prize for his study of the mechanism of chemical reactions; in 1958 three more Soviet scientists - P.A. Cherenkov, I.Ye. Tamm and I.M. Frank - also received a Nobel Prize for their discovery and interpretation of the "Cherenkov effect."

The Academy's chemists devised an industrial method of synthesizing isoprene, while a team of geologists under the direction of Academicians D.V. Nalivkin and N. S. Shatskiy compiled and published in 1956 and 1957 their 1:2,500,000 and 1:5,000,000 geological maps of the USSR and contiguous areas. In 1956 the celebrated helminthologist, Academician K.I. Skryabin, completed his 12-volume work *Trematody zhivotnykh i cheloveka* (Trematodes in Animals and Man) (by 1968 this had been extended to 22 volumes). Skryabin also received a Lenin Prize. In 1957 Academician A.N. Tupolev in turn received a Lenin Prize for designing the TU-104 and TU-114 passenger jets.

On 18 May 1957, in connection with the plan to mobilize the natural resources of the Soviet Union's Eastern regions, the USSR Council of Ministers decided to establish the Siberian Division of the USSR Academy of Sciences, with its center in Novosibirsk. By early 1959 the Siberian Division had 13 institutes and various other academic institutions. In addition, it administered the East Siberian, Far Eastern and Yakutsk Branches of the USSR Academy of Sciences, as well as the Buryat and Sakhalin Comprehensive Research Institutes and the Central Siberian Botanical Gardens. As early as March 1958 eight new full members and 27 corresponding members had been elected to the USSR Academy of Sciences from the Siberian Division.

Altogether from January 1956 to January 1959 the USSR Academy of Sciences expanded to the tune of 63 new scientific establishments, 48 of them institutes. In June 1958 some 26 new academicians and 55 corresponding members were elected. By 31 December 1959 the Academy had 158 full members, 344 corresponding members and a staff of over 15,000 persons.

In March 1959 the General Assembly of the USSR Academy of Sciences, upon instructions from the USSR Council of Ministers, approved the Academy's latest Statutes, reflecting the changes which had occurred in the Academy's work since adoption of the 1935 Statutes. The new Statutes stipulated that "The USSR Academy of Sciences carries out scientific research contributing to the development of all branches of science in the USSR. Through its research and all its work the Academy of Sciences plays an active part in the building of a Communist society in the USSR and helps to safeguard the socialist achievements of the workers and strengthen peace throughout the world."* The Statutes provided for the Academy to conduct its scientific work on the basis of strict planning: moreover, the plans it compiled had to be approved by the USSR Council of Ministers. The Statutes also confirmed the structure of the Siberian Division.

Between 1959 and 1961 the USSR Academy of Sciences was still further expanded with the founding of new institutions, including the African Institute in 1959 and the Latin American Institute in 1961. Most of the Siberian Division's research establishments were housed in new buildings at the Science Township near Novosibirsk. By 1961 the Division had 20 institutes and employed 3,000 specialists, including 36 academicians and corresponding members and more than 500 doctors and candidates of science. In May 1961 new presidential elections were held, and Academician M.V. Keldysh (the present incumbent) replaced A. N. Nesmeyanov as president of the USSR Academy of Sciences.

On 3 April 1961 the CPSU Central Committee and USSR Council of Ministers passed a joint resolution on "Measures to Improve the Coordination of Scientific Research Work in the Country and the Operation of the USSR Academy of Sciences," providing for further restructuring of the Academy of Sciences. It was now to

* *Nauchnyye kadry v SSSR. Sbornik dokumentov i spravochnykh materialov* (Scientific Personnel in the USSR. Collected Documents and Reference Material), Moscow, 1959, p 287.

concentrate on solving long-term scientific problems. In 1961, too, there was established the USSR Council of Ministers' special State Committee for the Coordination of Research, renamed in 1965 the State Committee for Science and Engineering. Ninety-two of the Academy of Sciences' scientific establishments, including all its branches, which were engaged in research connected with a particular branch of the national economy were removed from the Academy and turned over to the jurisdiction of state committees and other agencies. This was roughly half the Academy's scientific establishments; 20,500 of its scientific personnel were transferred with the institutions.

Many of these institutions and all the branches were later returned to the USSR Academy of Sciences, since "life had shown that the transfer of a number of establishments working on prospective fields of sciences was not dictated by objective factors."*

By the 11 April 1963 resolution of the CPSU Central Committee and USSR Council of Ministers on "Measures for Improving the Operation of the USSR Academy of Sciences and the Academies of Sciences of the Union Republics," the USSR Academy of Sciences was entrusted with the scientific direction of research on the major problems of the natural and social sciences conducted at the republican academies of sciences, higher educational establishments and other research institutions of the USSR. The Academy became the coordinating and guiding center of science in the USSR. In connection with this several reorganizational measures were taken and new Statutes approved on 1 July 1963, statutes which are still in force. Three sections - the Section of Physicotechnical and Mathematical Sciences, the Section of Chemicotechnological and Biological Sciences and the Section of Social Sciences - were established under the Presidium of the USSR Academy of Sciences to direct the 15 departments. In 1968 the Department of Earth Sciences was taken as the basis for a fourth section, the Section of Earth Sciences, which consisted of two departments: the Department of Geology, Geophysics and Geochemistry and the Department of Oceanology, Atmospheric Physics and Geography. In June

---

Akademiya nauk SSSR - shtab sovetskoy nauki (The USSR Academy of Sciences, Headquarters of Soviet Science), Moscow, 1968, p 162.

1964 some changes were made in the system of electing corresponding members and in the tasks of the Siberian Division.

Effective 1958, the election of new full and corresponding members was held every two years. In 1968 some 40 academicians and 80 corresponding members were elected, and by November 1970 the USSR Academy of Sciences had 237 full members, 445 corresponding members and 67 foreign members. Its total staff had also increased considerably. By early 1968 the Academy employed 25,000 scientific personnel, including 9,000 doctors and candidates of science.

In the last decade the USSR Academy of Sciences has expanded annually by the foundation of new institutes and other research establishments and by the return of several institutions which were earlier transferred to the jurisdiction of other agencies. Among the newly-founded institutes there appeared, for example, the Institute of the USA (1967), the Institute of Specific Social Research (1968), the Institute of Scientific Information on the Social Sciences (1969) and the Institute of the Far East (1969), which was founded in place of the previously abolished China Institute. In 1968 the USSR Academy of Sciences was given jurisdiction over the Institute of the International Workers' Movement and, also in 1968, the Institute of USSR History and the Institute of General History were formed from the Institute of History. The Institute of Asian Peoples was also renamed the Institute of Oriental Studies, and the Institute of Slavic Studies was renamed the Institute of Slavic and Balkan Studies. Many new institutes covering the natural sciences and technology were founded, but we have singled out the above-mentioned institutes specially. Like the Africa Institute (1969) and the Latin America Institute (1961), these institutes were founded for all-round study of the Asian, African and Latin American countries and the USA in order to facilitate the ideological penetration of these countries and promote the campaign against so-called "imperialist diversion."

In late 1969 and early 1970 the CPSU Central Committee and USSR Council of Ministers decided to organize major scientific centers in a number of the RSFSR's economically important areas. By the end of the current five-year plan such centers were scheduled for the Urals and Far East under the jurisdiction of the USSR Academy of Sciences. In the Far East the Academy of Sciences has 16 scientific institutions, including institutes: the Northeastern Comprehensive Research Institute, the Far Eastern Geological Institute, the Institute of Bioactive Substances, the Biological Soil Institute, the Sakhalin Comprehensive Research Institute, the Institute of Volcanology, the Institute of Marine Biology and the Khabarovsk Comprehensive Research Institute. These institutes and other scientific establishments are now being developed into the Far Eastern scientific center of the USSR Academy of Sciences. In addition, the USSR Academy of Sciences' Presidium has already drafted schedules and priorities for founding new establishments, has commissioned comprehensive measures for the development of research at the new scientific center and has approved the capital construction plan for 1971-75. Doctor of Geography A. P. Kapitsa has been appointed by the USSR Academy of Sciences' Presidium to oversee the Far Eastern Scientific Center.

In the past decade the Academy's scientists have made great headway in space research and the development of space rockets and spaceships. Back in 1959 three spaceships were launched toward the moon. The third of these, launched on 4 October, looped the moon, photographed its reverse side and transmitted these photographs back to the earth. On 12 February 1961 a space rocket was launched toward Venus from a heavy earth satellite. This rocket placed an automatic interplanetary station in a trajectory which intercepted with Venus. On 12 April 1961 man made his first space flight. In the Vostok-1 spaceship the Soviet Army major, Yu.A. Gagarin circled the earth in 108 minutes and landed in his target area in the USSR. On 6 and 7 August 1961 Major G. S. Titov made another space flight in Vostok-2, completing 17 orbits of the earth in the time of 25 hours.

The era of space flight and the conquest of space had dawned, involving studies of terrestrial space and the distant planets with the help of automatic interplanetary stations and devices. The Vostok-6 spaceship, launched on 16 June 1963, was piloted by the world's first woman astronaut V. V. Tereshkova, who spent 71 hours in space and completed 48 orbits of the earth. The first multi-seater spaceship, Voskhod-1, was launched into earth orbit on 12 October 1964. It carried the commander, V.M. Komarov, the research scientist K.P. Feoktistov and the physician B.B. Yegorov. The flight lasted 24 hours and involved 16 earth orbits. Voskhod-2 was launched into orbit on 18 March 1965, carrying Colonel P.I. Belyayev, the commander, and Lieutenant-Colonel A. A. Leonov as second pilot. During the flight Leonov emerged from the spaceship and spent several minutes in open space. On 14 and 15 January 1969 the Soyuz-4 and Soyuz-5 spaceships, piloted by V.A. Shatalov and B.V. Volynov, were launched into earth orbit. The ships also carried flight engineer A.S. Yeliseyev and research engineer Ye.V. Khrunov. In the course of the three-day orbital flight the world's first experimental space station was formed, thus testing automatic and manual control systems for the approach and docking of the two spaceships. Yeliseyev and Khrunov ventured into space and exchanged spaceships, spending about an hour in open space. In June 1970 the Soyuz-9 spaceship performed an 18-day orbital flight to test the possibility of establishing long-term orbital stations and the effects of prolonged space flight on the human organism. Soyuz-9 carried pilot-cosmonaut A.G. Nikolayev as commander and flight engineer V.I. Sevast'yanov, a research scientist. The program, which included many other tests and experiments, was fully and successfully carried out.

At the same time, Venus, Mercury, Mars and the moon were studied with the help of interplanetary stations and apparatus. The Venus-4 automatic station was launched on 18 October 1967 and, after a flight of four months, reached the planet of Venus and descended to its surface. During this descent much valuable data on the properties of near-Venetian space were obtained and the parameters of the Venetian atmosphere measured. Venus-7, launched on 17 August, reached Venus on 15 December 1970. During its 35-minute descent it transmitted valuable information back to earth before suddenly falling silent, evidently crushed or burned up in the planet's atmosphere. On 10 November 1970 Luna-17 started toward the moon from its orbit around the earth and on 17 November made a soft landing on the lunar surface in the area of the Sea of Rains. Luna-17 carried the self-propelled Lunokhod-1 moon-buggy, guided from the earth, which promptly emerged from the spacecraft and embarked on its research program. Prior to this, in September 1970, the moon was visited by the Luna-16 automatic station, which returned to earth on 24 September after taking core samples of the lunar surface.

Research conducted by the Academy of Sciences' physicists led to a start on the industrial production of artificial diamonds. The research on atomic energy and nuclear physics proceeded apace. New atomic reactors were put into operation, and as early as 1959 a Lenin Prize was awarded Academicians V. I. Veksler, A. L. Mints and their team for building a 10-Bev synchrophasotron. In 1964 plasma with a temperature of several dozen million degrees centigrade was obtained as a result of research directed by Academician G. I. Budker at the Kurchatov Institute of Atomic Energy and the Siberian Division's Institute of Nuclear Physics. In 1964 corresponding members of the USSR Academy of Sciences N.G. Basov and A.M. Prokhorov and the American scientist Taunus received a Nobel Prize for their work on quantum electronics. Also in 1964 a Lenin Prize was awarded to corresponding member B.M. Vul for his work on the development of semiconductor quantum generators.

Experimental research on the physics of elementary particles also made great strides. Experiments with the Serpukhov accelerator yielded unexpected results in research on the interaction of pi-mesons, k-mesons and anti-protons with protons and deutrons. The Kurchatov Institute of Atomic Energy achieved great progress in research on the physics of high-temperature plasma. Its "Tokamak" installation, consisting basically of closed magnetic traps, recorded neutrons of thermonuclear origin in quasi-stationary regimes. Thanks to the work of the Lebedev Physics Institute there developed considerable interest in the use of heavy-duty lasers for controlled thermonuclear reactions. The Vavilov Institute of Physical Problems managed to create a free-standing stationary pinch charge of high-temperature plasma with a high density level. The Institute of Crystallography developed new, efficient laser crystals for the steady generation of a laser beam at room temperature. Research at the Institute of Semiconductors and at Moscow University disclosed new anomally great magneto-optic effects in anti-ferromagnetics and ferrite-granites, pointing the way for the use of these materials in laser and computer engineering. The Topchiyev Institute of Petrochemical Synthesis developed a new class of high-molecular compounds called polysilicate-hydrocarbons which could be used for new methods of separating atmospheric gases and particularly for enriching the air with oxygen. The Protein Institute traced the chemical-primary-structure pattern in natural proteins, enabling their secondary structure (spirality) to be predicted with a probability of up to 80 percent.

In the field of mathematical logic important results were obtained by the Leningrad Branch of the Steklov Mathematics Institute which built a model of a system of equations with whole-number coefficients, solving the negative case of Gilbert's 10th problem, posed almost half a century ago. In geology the institutes of the USSR Academy of Sciences' Siberian Division studied the pattern for the distribution of placer and primary gold deposits in Siberia and the Far East, forecasting the location of such deposits. The USSR Academy of Science's Institute of Ore Deposit Geology, Petrography, Mineralogy and Geochemistry studied the distribution of platinum and platinoids in the ores of the Noril'sk-Talmakh deposits group. Of great importance in the prognostication and prospecting of minerals is the geological map of ancient formations compiled by the Institute of Geology and Geochronology of the Precambrian.

The process of de-Stalinization was also a great boon to the social sciences, particularly archeology and ethnography, cultural history, general history, linguistics, literary history, Oriental studies and anthropology. Many valuable studies, including works running to several volumes, appeared in these fields. In most cases those official brakes on the development of the social sciences, such as the prescriptions for invariably following the methodology of Marxism-Leninism, observing the principles of Lenin's Party primacy or promoting the struggle against the "ideological diversion of imperialism" were reflected only superficially. These Party prescriptions, however, still basically applied to Marxist-Leninist philosophy per se, the history of the CPSU, the history of the USSR, the history of international relations and works of a purely propaganda bent. For most Soviet sociologists, and particularly for the staff of the USSR Academy of Sciences, the period from 1968 to 1970 proved very barren. In connection with the 100th anniversary of

Lenin's birth, their efforts were channeled into "Leniniana" projects, involving the outpouring of a vast number of stereotyped books lauding Lenin's "unsurpassed" genius and the so-called modern Leninist policy of the CPSU leadership.

From 1955, with the promulgation of the "peaceful coexistence" policy, the USSR Academy of Sciences' international scientific contacts showed great development. In 1955 the Soviet Union joined UNESCO, while the USSR Academy of Sciences became a member of the International Council of Scientific Unions and, ipso facto, a member of all the major international scientific organizations which are members of this council. Russian was recognized as one of the three official languages of the International Council of Scientific Unions. Soviet scientists began to take an active and massive part not only in the work of international scientific conferences, congresses and gatherings but also in the work of various international scientific committees, commissions and expeditions. Many of these international scientific conferences, congresses and gatherings were held in the USSR. In addition, the USSR Academy of Sciences established direct scientific contacts and concluded scientific exchange agreements with academies of sciences and other scientific establishments and organizations in Europe, Asia, Africa, the USA and Canada. This was also accomplished via inter-governmental agreements which provided for joint seminars, exchanges of scientists and joint scientific projects of specialist training. Somewhat different in character were the USSR Academy of Sciences' scientific contacts with the academies of sciences and other scientific organizations of the so-called People's Democracies, for these contacts were on political grounds more active and all-embracing.

As the CPSU leadership constantly maintains, there can and should be no "peaceful coexistence" in ideology. The ideological struggle against capitalist countries and against "revisionism" in the socialist camp itself must continue to its victorious conclusion. This factor must be borne in mind in evaluating the scientific contacts which Soviet scientists and scholars develop with their counterparts in foreign countries. After all, these contacts also embody attempts to lull the vigilance of scientists and scholars in the free world and disarm or demoralize them ideologically, turning them into champions of the Soviet Union. These contacts are accompanied by a corresponding ideological campaign, as is proven by the numerous international congresses, gatherings and symposia called specially to discuss the social sciences. Soviet scientists and scholars, incidentally, head the various Soviet committees and societies for "friendship" and "solidarity" with countries in all continents; these are not, however, scientific bodies but purely political and propaganda organizations.

The structure and tasks of the USSR Academy of Sciences are defined by statutes and by the decisions of the USSR Council of Ministers and the General Assembly of the Academy itself. Its main task is, of course, to use its scientific establishments and personnel to organize scientific work and research of primary importance to the development of science, technology and the national economy. However, the USSR Academy of Sciences also has the function of coordinating and directing research throughout the USSR and is the main arbiter in determining the profile, general guidelines and specialization of all research establishments in the Soviet Union. The first step along these lines was taken back in 1945, when the Soviet government decided to establish under the USSR Academy of Sciences' Presidium a council to coordinate the scientific work of the union-republics' academies of sciences. In 1963 the USSR Academy of Sciences received extensive powers as the guiding, coordinating and controlling center of science in the USSR. These guiding and coordinating functions were later further expanded to such an extent that all research establishments are subordinate to it, no matter to which government agency they belong.

The USSR Academy of Sciences performs these functions by various ways and means. Firstly, on a national scale, it sets basic research guidelines and approves the annual and long-term scientific work plans submitted by other research establishments and, above all, the republican academies of sciences. In this context it also issued suggestions for the organization of research, material and technical supply and the development of the network of scientific establishments. Secondly, the USSR Academy of Sciences exercises its direction via numerous scientific problem councils. These councils bring together the country's leading scientists and specialists from various research establishments, draft long-term plans and examine individual problems in the natural and social sciences. Finally, the USSR Academy of Sciences arranges general and regional conferences, meetings and discussions to examine scientific achievements, adopt decisions and outline the plans and subjects of further research. The USSR Academy of Sciences cooperates with the USSR Council of Ministers' State Committee on the following basis: the Academy drafts and submits proposals for the annual and long-term state research plans and meets with the Committee to discuss and settle major scientific problems. Together with the Committee, the Academy of Sciences coordinates the development of academic establishments and helps to draw up annual and long-term research finance plans. All matters of principle affecting the organization of scientific research which are considered by the USSR Council of Ministers and other state organs are submitted for the USSR Academy of Sciences' conclusions or are debated together with Academy representatives.

The USSR Academy of Sciences has 16 departments, each a scientific and organizational center uniting establishments and scientists in one or several branches of science. There is also the Siberian Division, which incorporates members of the academy who are regularly employed at Siberian and Far Eastern establishments, regardless of their scientific speciality. As we have already noted, other scientific centers, such as the Far Eastern Scientific Center, are now being established. It is to be expected that they, too, will also eventually become divisions.

The legal status of a department is defined by the USSR Academy of Sciences' Statutes and the special "Regulations on Departments of the USSR Academy of Sciences." The supreme organ of a department is its general assembly. In the period between general assembly sessions the department is run by a Bureau, headed by the department's academic secretary. The department bureau is elected for a term of four years by the department's general assembly from members of the department staff and consists of the academic secretary, his deputies and other bureau members. One of the doctors or candidates of science proposed by the academic secretary is appointed learned secretary of the department by the USSR Academy of Sciences' Presidium and has the status of department bureau member. The department bureau is responsible to the department's general assembly and to the Presidium of the USSR Academy of Sciences. It submits an annual report on its work to the department's general assembly. Departmental decisions on matters within its jurisdiction do not require approval by the higher organ of the USSR Academy of Sciences. The department bureau directs its institutes and other scientific establishments, scientific councils, commissions, committees and periodicals which are part of this department.

The Siberian Division occupies a special position in the USSR Academy of Sciences. It incorporates the scientific establishments of the Academy situated in Siberia and the Far East. By its very nature it differs greatly from the Academy of Sciences' specialized departments. The Siberian Division is manned by full members and corresponding members of the Academy of Sciences, regardless of their speciality. Members of the Siberian Division enjoy all the rights of members of the department to which they belong according to their speciality. The Siberian Division of the USSR Academy of Sciences is subordinate not only to the USSR Academy of Sciences' Presidium but also to the RSFSR Council of Ministers. The Siberian Division's work plans are approved by the Presidium of the USSR Academy of Sciences in collaboration with the RSFSR Council of Ministers. The Siberian Division is headed by its chairman (at present Academician M.A. Lavrent'-yev), who is also a vice-president of the USSR Academy of Sciences.

The research institutes are the basic scientific organs of the USSR Academy of Sciences and are concentrated in the Academy's departments and branches. The structure, tasks and duties of the institutes and their organs are determined by the "Statutes of Scientific Research Institutes of the USSR Academy of Sciences" which are standard for all institutes. Overall direction of the departments and branch institutes is effected by the appropriate Sections of the USSR Academy of Sciences' Presidium. The departments are responsible for the scientific and organizational administration of their institutes. To which particular department an institute should belong is determined by the Presidium of the USSR Academy of Sciences. The institutes are structurally sub-divided into departments, sections and laboratories. An institute is headed by its director, elected for a term of four years from academicians, corresponding members and other eminent scientists by secret ballot at the general assembly of the respective department. His election is subject to confirmation by the general assembly of the USSR Academy of Sciences. The director is responsible for the institute's work to the general assembly of the department and the Academy of Sciences and also to the Presidium. To supervise the various aspects of an institute's work the director has several deputies: one or two on the science side and one for general or economic administrative work. Each institute has a Learned Council under the chairmanship of the director. The council is made up of the deputy director for scientific affairs, the learned secretary, the heads of departments, laboratories and sections and other scientists.

The USSR Academy of Sciences at present has 14 scientific societies. These societies incorporate scholars and scientists of one or several allied specialities. They are the: All-Union Astronomical and Geodesic Society; USSR Geographical Society; All-Union Mineralogical Society; All-Union Biochemical Society; All-Union Botanical Society; All-Union Hydrobiological Society; All-Union Society of Helminthologists; All-Union Microbiological Society; All-Union Society of Soil Scientists; All-Union Entomological Society; All-Union Pavlov Physiological Society; All-Union Vavilov Society of Geneticists and Selectionists; All-Union Society of Protizoologists; Russian Palestinian Society.

As a rule, each scientific society consists of full members and corresponding members of the society; some societies also have honorary members, elected from prominent scholars, scientists and public figures. These may be Soviet citizens or foreign researchers. Honorary members have the same rights as do full members of the society and are elected at the society's all-union congress. Full members, who may be researchers, practical specialists or persons whose work furthers the society's aims, are elected at general assemblies of the society's departments in the various cities of the USSR. There are also collective members of the society, i.e., other scientific societies, research institutes, higher educational establishments, technicums, secondary schools and various social organizations which promote the society's work. A society's highest organ is its all-union congress. In the period between congresses the society is administered by its Central Council, elected by the congress. The Presidium of the Central Council consists of the president, the vice-president, the learned secretary and members of the Presidium. Each society has its departments in the republics, autonomous republics, krays, oblasts and cities of the USSR. Overall direction of the Academy of Sciences' scientific societies is effected by the corresponding departments of the USSR Academy of Sciences.

The USSR Academy of Sciences has over 150 scientific councils on various problems of science. These incorporate the country's leading scientists, whose function is to evaluate the state of research in their field and channel it along optimum lines. The USSR Academy of Sciences has scientific councils covering problems in the natural and social sciences; the scientific councils on scientific and technical problems come under the USSR Council of Ministers' State Committee for Science and Engineering. The scientific councils of the USSR Academy of Sciences are established by its Presidium. These councils include representatives of the establishments of the USSR Academy of Sciences, the republican academies of sciences, branch academies and institutes and the scientific establishments of universities. One of the scientific councils' main tasks is to draft annual and long-term research plans for the country's scientific establishments. They also engage in research proper, undertaking major projects on individual problems. In fulfillment of this task, the scientific councils are also coordinating organs and enjoy quite extensive powers. The scientific councils draft proposals for the development of further research and the incorporation of research results in the national economy and submit these proposals for approval by the appropriate agencies and organizations.

In addition to the USSR Academy of Sciences' scientific councils, there are over 260 scientific councils at the republican academies of sciences. For the most part, these are but territorial sections of the USSR Academy of Sciences' scientific councils, but some of them do have all-union status. There are also a number of regional scientific councils, i.e., councils for problems covered by two or more republican academies of sciences. Scientific councils are restricted to general and comprehensive problems.

In addition to the USSR Academy of Sciences and the academies of sciences of the Soviet national republics (14 in all; the RSFSR does not have its own individual academy of sciences; the Moldavian Academy of Sciences was founded in 1960), the Soviet Union also has branch academies which come under the jurisdiction of the appropriate ministries. These are the USSR Academy of Medical Sciences (founded 1944), the All-Union Lenin Academy of Agricultural Sciences (1929), the USSR Academy of Arts (1947) and the USSR Academy of Pedagogic Sciences, founded in 1967 from the old RSFSR Academy of Pedagogic Sciences. All these operate in close cooperation with the USSR Academy of Sciences.

The present system for electing full and correspinding members of the USSR Academy of Sciences is set forth in its 1 July 1963 Statutes and 21 June 1964 "Regulations for Elections to the USSR Academy of Sciences." (The Regulations' clause on maximum age of candidates for corresponding membership was deleted by the General Assembly of the USSR Academy of Sciences in March 1968.) Elections are held at least once every two years by the General Assembly of the USSR Academy of Sciences in secret ballot. Only full members of the Academy are eligible to vote. To be elected a candidate must poll at least two-thirds of the votes of all full members. Many candidates, of course, become members of the Academy through Party protection, involving manipulation of the elections.

Every member of the USSR Academy of Sciences, in accordance with his speciality, belongs to one of its departments, in which he enjoys the rights established for members of the Academy. Members of the Academy may belong simultaneously to several departments, in which case they may wield their vote in elections to the Academy of Sciences in only one department.

Corresponding members of the USSR Academy of Sciences have voting rights at general assemblies of the departments on all matters under consideration with the exception of the election of members of the Academy. In the General Assembly of the USSR Academy of Sciences they enjoy a deliberative voice. They are also eligible for election to the executive organs of the departments, as directors of the departments' scientific institutions and to fill vacant posts outside the normal competition.

The General Assembly is the USSR Academy of Sciences' highest organ and meets in session as and when required, but not less than twice a year. Sessions last two or three days. Since 1968 the annual General Assembly session to summarize the Academy's work for the past year has been held in the first ten days of March. The General Assembly takes decisions by a simple majority vote, except for those cases where the Statutes require a qualified majority of two-thirds of the votes (the election of members of the USSR Academy of Sciences) or three-quarters (the election of a Presidium member for more than two straight terms of office). The General Assembly confirms the Statutes of the USSR Academy of Sciences, as well as the "Regulations on Elections to the USSR Academy of Sciences," "Regulations on Departments of the USSR Academy of Sciences," "Regulations on Branches of the USSR Academy of Sciences" and "The Statutes of Scientific Research Institutes of the USSR Academy of Sciences."

Between General Assembly sessions the Academy of Sciences is run by the Presidium of the USSR Academy of Sciences, consisting of the president, vice-presidents, the chief learned secretary of the Presidium, the academic secretaries of departments and Presidium members, to a number specified by the General Assembly of the USSR Academy of Sciences.

The Presidium is elected from full members of the Academy for a term of four years. Elections are by means of secret ballot with a simple majority of full members' votes. The chief learned secretary of the Presidium is elected by the Presidium itself. As a rule, members of the Presidium are not elected for more than two successive terms.

From among its members the Presidium forms Presidium sections responsible for the corresponding departments and scientific establishments of the Academy. At present there are four sections: Physicotechnical and Mathematical Sciences, Earth Sciences (from

1968), Chemicotechnological and Biological Sciences, and Social Sciences. The sections are manned by the academic secretaries of departments which come under the section and by Presidium members of appropriate specialities. The sections are headed by chairmen.

The Presidium's work is shared among the president, the vice-presidents and the other Presidium members, and neither the president, the vice-presidents nor the chief learned secretary of the Presidium enjoy any special rights under the Statutes. They are responsible individually to the Presidium, and the Presidium is collectively responsible to the General Assembly of the USSR Academy of Sciences.

When the Presidium considers matters connected with the work of the republican academies of sciences, the presidents of the republican academies of sciences are entitled to participate as voting delegates in the meeting of the USSR Academy of Sciences' Presidium. The USSR Academy of Sciences Presidium's Council for the Coordination of the Scientific Work of the Union-Republic Academies of Sciences is headed by the president of the USSR Academy of Sciences and consists of the presidents of the republican academies of sciences, the academic secretaries of the USSR Academy of Sciences' departments, the chairman of the USSR Council of Ministers' State Committee for Science and Engineering, the chairman of the presidium of the USSR Academy of Sciences' branches, the president of the USSR Academy of Medical Sciences, the president of the All-Union Lenin Academy of Agricultural Sciences and the USSR Minister of Higher and Secondary Specialized Education.

New recruits to staff the USSR Academy of Sciences' institutions are drawn mainly from the Academy's postgraduate training system, which turns out candidates and doctors of science. Prominent scientists and scholars from the universities and other scientific establishments are also recruited to work at the Academy. For outstanding scientific works and discoveries the Presidium of the USSR Academy of Sciences confers on individual scientists gold medals and various name prizes. Most of these medals and prizes are awarded once every three years.

In the past decade the USSR Academy of Sciences

has paid considerable attention to a more efficient distribution and location of its scientific establishments. Whereas in 1951 some 90 percent of its institutes were concentrated in Moscow and Leningrad, by 1966 only 65 percent of its scientific institutions were located in these cities. In early 1966 more than 70 of the Academy's institutes were sited in various economic areas of the USSR, especially in the Eastern regions of the RSFSR.

The USSR Academy of Sciences and its institutions arrange scientific sessions, meetings, conferences and symposia to debate individual problems or discuss the achievements and prospects for the development of scientific research. More than 300 such sessions, conferences and meetings are held every year.

The USSR Academy of Sciences' editing and publishing work is planned and coordinated by the USSR Academy of Sciences' Editing and Publishing Council, headed by the first vice-president of the USSR Academy of Sciences (presently Academician M. D. Millionshchikov) and his two deputies (presently vice-president A. M. Rumyantsev and Academician A. L. Yanshin). The council also includes leading scientists of all specialities, the chairmen of the republican academies of sciences' editorial councils, the academic secretaries of USSR Academy of Sciences' departments, the chairmen of the editorial boards of a number of serial academic publications, executive officials of the "Science" Press and representatives of the USSR Council of Ministers' Press Committee.

The USSR Academy of Sciences' publishing house was formally founded in 1924. Prior to this publishing work had been handled by the Academy's Chancellery and by Petersburg's oldest printing works, which belonged to the Academy. When the Academy of Sciences moved from Leningrad to Moscow, the publishing house moved with it. In 1963 the USSR Academy of Sciences' Publishing House was reorganized as the "Science" Press. It is the largest scientific publishing house in the Soviet Union. The Publishing House currently has 20 branch book-publishing boards, the Main Editorial Board for Physics and Mathematics Literature, the Main Editorial Board for Oriental Literature and a large periodical publications department with 120 journal editorial boards. The publishing house has its departments in Leningrad and Novosibirsk and operates three printing works, two in Moscow and one in Leningrad. Since the war there has been a great increase in the number of works published by the USSR Academy of Sciences. In 1940 over 12,000 titles of books and journals were published, but then the war radically cut back the Academy's publishing work. In 1943 its output was less than 2,000 titles. From 3,000 titles of books in 1945 the output increased to 31,600 book titles in 1966. In the same period the output of journals rose from 2,500 to 15,000 titles. In addition to its wide range of scientific publications, the Academy of Sciences has also published such collections of the Russian literary classics as the series "Classics of Literature" and "Literary Monuments," as well as popular-science series.

The publication of abstracts journals, bibliographical reviews and reference literature, as well as the translation of foreign literature and microfilming are assigned to the USSR Academy of Sciences' All-Union Institute of Scientific and Technical Information and the USSR Council of Ministers' State Committee for Science and Engineering. Since 1968 these tasks have also been handled by the Academy's newly-founded Institute of Scientific Information on the Social Sciences. At present some 187 editions of abstracts journals are published, constituting the Soviet Union's most complete collection of periodical data on science and engineering.

The USSR Academy of Sciences and its institutions are financed from the state budget. The budget levels for some of the years since the war were as follows (discounting the institutions of the USSR Academy of Sciences' Siberian Division; expressed in millions of rubles and equated to 1961 prices): 1946 - 48.7; 1951 - 74.6; 1955 - 122.9; 1961 - 129.2; 1965 - 152.2.

The USSR Academy of Sciences' institutions submit to the Academy's Presidium annual estimates of the funds needed for the fulfillment of their research plan in the coming year. After the state budget has been approved the USSR Academy of Sciences' Presidium apportions the funds assigned to the Academy among its various sections and departments in accordance with their future work plans. These funds are broken down into the following headings: research work, the acquisition of scientific equipment, and the staff wage and salary fund (with a limit for each department). The Presidium's sections and the departments distribute their portion of the funds among the scientific establishments according to the set limits. If an establishment finds it necessary during the course of its work to increase its staff and, hence, its wage fund, a proposal to this effect is submitted to the Presidium of the USSR Academy of Sciences.

The Section of Physicotechnical and Mathematical Sciences receives about 42 percent of the Academy of Sciences' annual budget. Most of this money is spent on purely theoretical research. The institutions which come under the Section of Chemicotechnological and Biological Sciences spend about 28 percent of the budget, distributed more or less evenly between theoretical and applied research. The Section of Social Sciences receives 10 percent of the Academy's annual budget. The remaining 20 percent is spent on the academy as a whole to cover such needs as scientific information, the operation of libraries and the work of the various commissions and communal academic establishments.

Chapter II

HEADS OF THE USSR ACADEMY OF SCIENCES

Presidium of the USSR Academy of Sciences

PRESIDENTS OF THE USSR ACADEMY OF SCIENCES

Karpinskiy, A.P.	1917-1936
Komarov, V.L.	1936-1945
Vavilov, S.I.	1945-1951
Nesmeyanov, A.M.	1951-1961
Keldysh, M.V.	1961-

VICE-PRESIDENTS OF THE USSR ACADEMY OF SCIENCES

Borodin, I.P.	1917-1919
Steklov, V.A.	1919-1926
Fersman, A.Ye.	1927- ?
Marr, N.Ya.	1930-1934
Komarov, V.L.	1930-1936
Krzhizhanovskiy, G.M.	1930-1939
Britske, E.V.	1936-1939
Gubkin, I.M.	1936-1939
Chudakov, Ye.A.	1936-1939
Shmidt, O.Yu.	1938-1942
Bogomolets, A.A.	1942-1944
Orbeli, L.A.	1942-1946
Volgin, V.P.	1942-1953
Bardin, A.P.	1942-1960
Ostrovityanov, K.V.	1953-1962
Lavrent'yev, M.A.	1957-
Topchiyev, A.V.	1958-1962
Keldysh, M.V.	1960-1961
Kirillin, V.A.	1962-1965
Millionshchikov, M.D.	1962-
Semyonov, N.N.	1962-
Fedoseyev, P.N.	1964-1966
Konstantinov, B.P.	1966-1969
Vinogradov, A.P.	1967-
Rumyantsev, A.M.	1967-

SECRETARIES OF THE USSR ACADEMY OF SCIENCES' PRESIDIUM

Permanent Secretaries

Ol'denburg, S.F. 1904-1929
Volgin, V.P. 1929-1935
Gorbunov, N.P. 1935-1938

Chief Learned Secretaries

Bruyevich, N.G. . . . . ? -1949
Topchiyev, A.V. 1949-1958
Fyodorov, Ye.K. 1959-1962
Sisakyan, N.M. 1963-1966
Peyve, Ya.V. 1966-

Presidium Members
(Academic Secretaries of the USSR Academy of Sciences' Departments)

Shakhmatov, A.A. 1917-1920
Sushkin, P.P. 1927-1928
Borisyak, A.A. 1934-1935
Orlov, A.S. 1934-1935
Fersman, A.Ye. 1935-   ?
Grebenshchikov, I.V. 1934-   ?
Shenfer, K.I. 1934-1936
Grekov, B.D. 1935-1938
Britske, E.V. 1936-1938
Stepanov, P.I. 1938-1941
Bakh, A.N. 1938-1945
Ioffe, A.F. 1938-1949
Meshchaninov, I.I. 1938-1951
Nikitin, V.P. 1939-1942
Orbeli, L.A. 1939-1948
Obruchev, V.A. 1942-1946
Vvedenskiy, B.A. 1944-1953
Grekov, B.D. 1944-1953
Nesmeyanov, A.N. 1946-1948
                                               1963-
Zavaritskiy, A.N. 1947-1949
Dubinin, M.M. 1948-1957
Petrovskiy, I.G. 1949-1951
Belyankin, D.S. 1949-1953
Ostrovityanov, K.V. 1949-1953
Oparin, A.I. 1949-1956
Lavrent'yev, M.A. 1951-1957
Vinogradov, V.V. 1951-1963

Blagonravov, A.A. 1953
                                           1957-1963
Khristianovich, S.A. 1953-1956
Tikhomirov, M.N. 1953-1957
Shcherbakov, D.I. 1953-1963
Nemchinov, V.S. 1954-1959
Engel'gard, V.A. 1956-1959
Semyonov, N.N. 1957-1963
Zhukov, Ye.M. 1957-
Artsimovich, L.A. 1957-
Fedoseyev, P.N. 1959-1967
Sisakyan, N.M. 1959-1963
Arzumanyan, A.A. 1962-1965
Bogolyubov, N.N. 1963-
Veksler, V.I. 1963-1966
Millionshchikov, M.D. 1963-1965
Vinogradov, A.P. 1963-1967
Petrov, B.N. 1963-
Zhavoronkov, N.M. 1963-
Shemyakin, M.M. 1963-1970
Chernigovskiy, V.N. 1963-1967
Bykhovskiy, B.Ye. 1963-
Khrapchenko, M.B. 1963-
Rumyantsev, A.M. 1965-1967
Styrikovich, M.A. 1965-
Khachaturov, T.S. 1967-
Markov, M.A. 1967-
Sadovskiy, M.A. 1967-
Kreps, Ye.M. 1967-
Konstantinov, F.V. 1967-

Presidium Members

Frumkin, A.N. 1936-1939
Britske, E.V. . . . . 1939- ?
Obraztsov, V.N. 1939-1945
Chudakov, Ye.A. . . . . 1942- ?
Muskhelishvili, N.I. 1941-
Khristianovich, S.A. 1946-1953
                                           1957-1962
Satpayev, K.I. 1946-1952
                                           1955-1964
Palladin, A.V. 1946-1962
Grebenshchikov, I.V. . . . . 1947- ?
Petrovskiy, I.G. 1953-
Abartsumyan, V.A. . . . . 1953- ?
Keldysh, M.V. 1953-1960
Kapitsa, P.L. 1955-

Kursanov, A.L. 1955-1962
Kurchatov, I.V. 1957-1960
Dubinin, M.M. 1957-1962
Arbuzov, A.Ye. . . . . ? -1962
Kostenko, M.P. 1957-1962
Nemchinov, V.S. 1960-1962
Aleksandrov, A.P. 1960-
Yudin, P.F. 1960-1962
Nesmeyanov, A.N. 1960-1963
Korolyov, S.P. 1960-1966
Paton, B.Ye. 1962-
Trofimuk, A.A. 1963-
Pospelov, P.N. 1963-
Kotel'nikov, V.A. 1963-
Chelomey, V.N. 1963-1965
Basov, N.G. 1966-
Mel'nikov, N.V. 1966-
Pilyugin, N.A. 1966-
Fedoseyev, P.N. 1967-

Departments of the USSR Academy of Sciences

ACADEMIC SECRETARIES

Under the 1836 Statutes:

* 1 Department of Physical and Mathematical Sciences
        1917-1927

  ? 1917-1927

* 2 Department of History and Philology
        1917-1927

  ? 1917-1927

* 3 Department of Russian Language and Literature
        1917-1927

  Shakhmatov, A.A. 1917-1920
  ? 1920-1927

Under the 1927 Statutes:

* 4   Department of Mathematical and Natural Sciences
              1927-1935

       Sushkin, P.P. 1927-1928
       ?   1928-1934
       Borisyak, A.A. 1934-1935

* 5   Department of Humanitarian Sciences
              1927-1935

       ?   1927-1934
       Orlov, A.S. 1934-1935

Under the 1935 Statutes:

* 6   Department of Mathematical and Natural Sciences
              1935-1938

       Fersman, A.Ye. 1935-1938

* 7   Department of Technical Sciences
              1935-1938

       Grebenshchikov, I.V. 1935
       Shenfer, K.I. 1935-1936
       Britske, E.V. 1936-1938

* 8   Department of Social Sciences
              1935-1938

       Grekov, B.D. 1935-1938

After 1938 reorganization:

* 9   Department of Physical and Mathematical Sciences
              1938-1963

       Ioffe, A.F. 1938-1949

      Petrovskiy, I.G. 1949-1951
      Lavrent'yev, M.A. 1951-1957
      Artsimovich, L.A. 1957-1963

* 10 Department of Chemical Sciences
      1938-1963

      Bakh, A.N. 1938-1945
      Nesmeyanov, A.N. 1946-1948
      Dubinin, M.M. 1948-1957
      Semyonov, N.N. 1957-1963

* 11 Department of Geology and Geography
      1938-1963

      Stepanov, P.I. 1938-1941
      Obruchev, V.A. 1942-1946
      Zavaritskiy, A.N. 1947-1949
      Belyankin, D.S. 1949-1953
      Shcherbakov, D.I. 1953-1963

* 12 Department of Biology
      1938-1963

      Orbelli, L.A. 1938-1948
      Oparin, A.I. 1949-1956
      Engel'gard, V.A. 1956-1959
      Sisakyan, N.M. 1959-1963

* 13 Department of Technical Sciences
      1938-1963

      ? 1938
      Nikitin, V.P. 1939-1942
      ? 1942-1944
      Vvedenskiy, B.A. 1944-1953
      Blagonravov, A.A. 1953
      Khristianovich, S.A. 1953-1956
      Blagonravov, A.A. 1957-1963

* 14   Department of History and Philosophy
             1938-1953

         ? 1938-1944
         Grekov, B.D. 1944-1953

* 15   Department of Historical Sciences
             1953-1963

         Tikhomirov, M.N. 1953-1957
         Zhukov, Ye.M. 1957-1963

* 16   Department of Economics and Law
             1938-1953

         ? 1938-1948
         Ostrovityanov, K.V. 1949-1953

* 17   Department of Economics, Philosophy and Law
             1953-1962

         Nemchinov, V.S. 1953-1959
         Fedoseyev, P.N. 1959-1962

* 18   Department of Philosophy and Law
             1962-1963

         Fedoseyev, P.N. 1962

* 19   Department of Economics
             1962-1963

         Arzumanyan, A.A. 1962-1963

* 20   Department of Literature and Language
             1938-1963

         Meshchaninov, I.I. 1938-1951
         Vinogradov, V.V. 1951-1963

Under the 1963 Statutes:

By the new statutes of 1 July 1963 there were founded 15 Departments of the USSR Academy of Sciences' Presidium, organized into three Sections, each of the latter headed by a vice-president of the Academy. On 7 March 1968 a fourth Section of the USSR Academy of Sciences' Presidium was established.

I. *PHYSICOTECHNICAL AND MATHEMATICAL SCIENCES SECTION OF THE USSR ACADEMY OF SCIENCES' PRESIDIUM*

1963-

21 Department of Mathematics
1963-

    Bogolyubov, N.N. 1963-

22 Department of General and Applied Physics
1963-1968

    Artsimovich, L.A. 1963-1968

23 Department of General Physics and Astronomy
1968-

    Artsimovich, L.A. 1968-

24 Department of Nuclear Physics
1963-

    Veksler, V.I. 1963-1966
    Markov, M.A. 1967-

25 Department of Physicotechnical Power Engineering Problems
1963-

    Millionshchikov, M.D. 1963-1965
    Styrikovich, M.A. 1965-

* 26   Department of Earth Sciences
         1963-1968

   Vinogradov, A.P. 1963-1967
   Sadovskiy, M.A. 1967-1968

27   Department of Mechanics and Control Processes
         1963-

   Petrov, B.N. 1963-

*II. EARTH SCIENCES SECTION OF THE USSR ACADEMY OF
     SCIENCES' PRESIDIUM*

         1968-

28   Department of Geology, Geophysics and Geochemistry
         1968-

   ?   1968-

29   Department of Oceanology, Atmospheric Physics and Geography
         1968-

   ?   1968-

III. CHEMOTECHNOLOGICAL AND BIOLOGICAL SCIENCES SECTION OF THE USSR ACADEMY OF SCIENCES' PRESIDIUM

         1963-

30   Department of General and Technical Chemistry
         1963-

   Nesmeyanov, A.N. 1963-

31 Department of Physicochemistry and Inorganic Materials Technology
1963-

   Zhavoronkov, N.M. 1963-

32 Department of Biochemistry, Biophysics and Active Physiological Compounds Chemistry
1963-

   Shemyakin, M.M. 1963-1970
   Belozerskiy, A.N. 1970-

33 Department of Physiology
1963-

   Chernigovskiy, V.N. 1963-1967
   Kreps, Ye.M. 1967-

34 Department of General Biology
1963-

   Bykhovskiy, B.Ye. 1963-

IV. SOCIAL SCIENCES SECTION OF THE USSR ACADEMY OF SCIENCES' PRESIDIUM

1963-

35 Department of History
1963-

   Zhukov, Ye.M. 1963-

36 Department of Philosophy and Law
1963-

   Fedoseyev, P.N. 1963-1967
   Konstantinov, F.V. 1967-

37 Department of Literature and Language
            1963-

        ? 1963-1966
        Khrapchenko, M.B. 1966-

38 Department of Economics
            1963-

        Arzumanyan, A.A. 1963-1965
        Rumyantsev, A.M. 1965-1967
        Khachaturov, T.S. 1967-

Institutions of the USSR Academy of Sciences' Presidium

* 39 All-Union Institute of Scientific and Technical
      Information
            1952-

            Directors

        Panov, D.Yu. 1952-1956
        Mikhaylov, A.I. 1956-1970
        Delyusin, L.P. 1970-

* 40 Library of the USSR Academy of Sciences
            1714-

            Directors

        ? 1917-1940
        Yakovkin, I.I. . . . . 1940- ?
        Chebotaryov, G.A. 1958-1960
        Filippov, M.S. 1960-1968
        Moiseyeva, A.A. 1968-1970
        Ter-Avanesyan, D.V. 1970-

41 "Science" Publishing House
      (Formerly called: Publishing House of the USSR
      Academy of Sciences)
            1924-

Directors

?	1924-1933
Kuz'min, G.N.	? -1947
Nazarov, A.I.	1947-1954
Morozov, A.V.	1954-1958
Nazarov, A.I.	1958-1962
Samsonov, A.M.	1962-1970
Komkov, G.D.	1970-

* 42   Editorial and Publishing Council
             ? -

Chairmen

Komarov, V.L.	1936-1945
Vavilov, S.I.	1945-1951
Nesmeyanov, A.N.	1951-1961
Keldysh, M.V.	1961- ?
Millionshchikov, M.D.	? -

* 43   Commission for the Study of Natural Production Resources of (Russia) USSR
             1915-1930

Chairman

Vernadskiy, V.I.	1915-1930

* 44   Council for the Study of Production Resources of USSR
             1930-1960

Chairmen

Gubkin, I.M.	1930-1936
Komarov, V.L.	1938-1945
Shevyakov, L.D.	1946-1949
Nemchinov, V.S.	1949-1960

  45   Commission for the Study of Production Resources and Means
             1967-

Chairman

Mel'nikov, N.V. 1967-

46 Joint Scientific Council for Physics and Chemistry of Semiconductors
1968-

Chairman

Vul, B.M. 1968-

47 Museum Council
1966-

Chairman

Rybakov, B.A. 1966-

48 Scientific Council of Ural Institutions
1968-

Chairman

Vonsovskiy, S.V. 1968-

49 Interdepartmental Council for Seismology and Seismostable Construction
1967-

Chairman

Savarenskiy, Ye.F. 1967-

50 Scientific Council for Cybernetics
1963-

Chairman

? 1963-

51 Scientific Council for Philosophical Problems of
   Modern Natural Sciences
            1959-

Chairman

Fedoseyev, P.N. 1959-

Institutions of the USSR Academy of Sciences

I. *PHYSICOTECHNICAL AND MATHEMATICAL
        SCIENCES SECTION*

* 52 Physics Laboratory
            1917-1921

Director

Krylov, A.N. 1917-1921

* 53 Institute of Physics and Mathematics
            1921-1934

Directors

Steklov, V.A. 1921-1926
Krylov, A.N. 1927-1932
Vinogradov, I.M. 1932-1934

54 V.A. Steklov Mathematics Institute
            1934-

Director

Vinogradov, I.M. 1934-

55 P.N. Lebedev Physics Institute
            1934-

Directors

    Vavilov, S.I. 1932-1951
    Skobel'tsyn, D.V. 1951-

* 56 Acoustics Commission
              ? -1938

Chairman

    Andreyev, N.N. . . . . ? -1938

  57 Computer Center
              1954-

Director

    Dorodnitsyn, A.A. 1954-

* 58 Leningrad Computer Center
              1959-

Director

    Posnov, N.N. 1959-

  59 Commission for Computer Engineering
              1957-

Chairman

    Dorodnitsyn, A.A. 1957-

* 60 Sverdlovsk Institute of Mathematics and Mechanics
              1970-

Director

    Krasovskiy, N.N. 1970-

61 Institute of Theoretical and Experimental
    Physics
                    1958-1964

                    Director

    Alikhanov, A.I. 1958-1964

62 L.D. Landau Institute of Theoretical Physics
                    1964-

                    Directors

    Alikhanov, A.I. 1964-1966
    Khalatnikov, I.M. 1966-

* 63 Acoustics Institute
                    1953-1961

                    Director

    Brekhovskikh, L.M. 1953-1961

64 Institute of Crystallography
                    1944-

                    Directors

    Shubnikov, A.V. 1944-1962
    Vaynshteyn, B.K. 1962-

* 65 Laboratory of High-Pressure Physics
                    1954-1958

                    Director

    Vereshchagin, L.F. 1954-1958

66 Institute of High-Pressure Physics
                    1958-

Director

Vereshchagin, L.F. 1958-

* 67   Institute of Metal Physics
         1958-

Director

Mikheyev, M.N. 1958-

68   Institute of Solid-State Physics
      1963-

Director

Kurdyumov, G.V. 1963-

69   Institute of Physical Problems
      1935-

Directors

Kapitsa, L.P. 1935-1946
Aleksandrov, A.P. 1946-1954
Kapitsa, L.P. 1954-

70   Permanent Atomic Nucleus Commission
         1938-   ?

Chairman

Vavilov, S.I. 1938-   ?

* 71   Institute of Nuclear Problems
         ?   -1956

Director

Dzhelepov, V.P. 1948-1956

* 72   Institute of Atomic Energy
            ?   -1963

   Directors

   Kurchatov, I.V. . . . . ? -1960
   Aleksandrov, A.P. 1960-1963

* 73   V.G. Khlopin Radium Institute
            1938-1961

   Directors

   Vernadskiy, V.I. 1938-1939
   Khlopin, V.G. 1939-1950
   Vdovenko, V.M. 1950-1960

   74   Institute of Radio Engineering and Electronics
            1953-

   Director

   Kotel'nikov, V.A. 1953-

* 75   Radio Engineering Institute
            1957-1961

   Director

   Mints, A.L. 1958-1961

* 76   A.F. Ioffe Institute of Physical Engineering
            1918-

   Directors

   Ioffe, A.F. 1918-1950
   Komar, A.P. 1950-1957
   Konstantinov, B.P. 1957-1967
   Tuchkevich, V.M. 1967-

* 77   Kazan' Institute of Physical Engineering
            ?   -

Director

    Mushtari, Kh.M. 1963-

* 78 G.M. Krzhizhanovskiy Power Engineering Institute
       1930-1961

Directors

    Krzhizhanovskiy, G.M. 1930-1959
    Kruzhilin, G.N. 1960-   ?

* 79 Semiconductors Laboratory
       1953-1954

Director

    Ioffe, A.F. 1953-1954

  80 Semiconductors Institute
       1954-

Directors

    Ioffe, A.F. 1954-1960
    ? 1960-1962
    Regel', A.R. 1962-

* 81 High Temperatures Laboratory
       1960-1961

Directors

    Kirillin, V.A. 1960-1961
    Sheydlin, A.Ye. 1961

* 82 High Temperatures Institute
       1967-

Director

    Sheydlin, A.Ye. 1967-

83 Laboratory of High-Frequency Electrometry
1947-

Director

Vologdin, V.P. 1947-

* 84 Institute of Comprehensive Transport Problems
1954-1960

Directors

Khachaturov, T.S. 1954-1959
Belousov, I.I. 1959-1960

85 Permanent Commission for Scientific Problems of Transport Development
1965-

Chairman

Gorinov, A.V. 1965-

* 86 Institute of Machine Science
1938-1961

Directors

Chudakov, Ye.A. 1938-1953
Blagonravov, A.A. 1953-1961

* 87 Engine Laboratory
1953-1961

Director

Stechkin, B.S. 1953-1961

* 88 Laboratory of Electric Welding Machines
1955-1961

Director

    Nikitin, V.P. 1955-1961

* 89 Laboratory of Control Machinery and Systems
         ?   -1958

Director

    Bruk, I.S. 1956-1958

* 90 Institute of Electronic Control Machinery
       1958-1961

Director

    Bruk, I.S. 1958-1961

* 91 Commission for Telemechanics and Automation
       1934-1938

Chairman

    Kulebakin, V.S. 1934-1938

  92 Committee for Automation
       1938-1961

Chairman

    Kulebakin, V.S. 1938-1961

  93 Institute of Automation and Telemechanics
       1938-

Directors

    Kovalenkov, V.I. 1938-1948
    Petrov, B.N. 1948-1953
    Trapeznikov, V.A. 1953-

* 94 Institute of Electromechanics
           1956-1961

           Director

       Kostenko, M.P. 1956-1961

* 95 Laboratory for the Scientific Development of
    Wire Communications Problems
           1948-1959

           Directors

       Kovalenkov, V.I. 1948-1957
       Kharkevich, A.A. 1957-1959

* 96 Laboratory of Data Transmission Systems
           1959-1962

           Director

       Kharkevich, A.A. 1959-1962

  97 Institute of Data Transmission Problems
           1962-

           Directors

       Kharkevich, A.A. 1962-1966
       Siforov, V.I. 1966-

* 98 Institute of Fine Mechanics and Computer
    Engineering
           1946-1961

           Directors

       Lavrent'yev, M.A. 1948-1953
       Lebedev, S.A. 1953-1961

* 99 Institute of Mechanics
           1939-1964

Directors

Chetayev, N.G.	? -1953
Il'yushin, A.A.	1953-1960
Nikol'skiy, A.A.	1960-1964

100 Institute of Mechanical Problems
     1964-

Director

Ishlinskiy, A.Yu. 1964-

* 101 Main Astronomical Observatory
      1934-

Directors

Gerasimovich, B.P.	1934-1937
Belyavskiy, S.I.	1937-1944
?	1944-1947
Mikhaylov, A.A.	1947-1965
Krat, V.A.	1965-

* 102 Commission for Solar Research
      1932-1934

Chairman

Gerasimovich, B.P. 1932-1934

103 Astronomical Council
    1937-

Chairmen

Mikhaylov, A.A.	1939-1963
Mustel', E.R.	1963-

* 104 Crimean Astrophysical Observatory
      1945-

Directors

 Shayn, G.A. 1945-1952
 Severnyy, A.B. 1952-

105 Special Astrophysical Observatory
   1966-

Director

 Kopylov, I.M. 1966-

* 106 Astronomical Institute
   1939-1943

Directors

 ? 1939-1942
 Subbotin, M.F. 1942-1943

107 Institute of Theoretical Astronomy
   1943-

Directors

 Subbotin, M.F. 1943-1964
 Chebotaryov, G.A. 1964-

* 108 Institute of Earth Magnetism, Ionosphere and Propagation of Radio Waves
   1959-

Directors

 Pushkov, N.V. 1960-1969
 Migulin, V.V. 1969-

109 Institute of Space Research
   1968-

Director

 Petrov, G.I. 1968-

* 110 Spectroscopic Commission
1940-

Chairmen

Landsberg, G.S. 1940-1957
Mandel'shtam, S.L. 1967-

111 Institute of Spectroscopy
1969-

Director

Mandel'shtam, S.L. 1969-

112 Scientific Council for the Use of Computer Equipment and Means of Automation in Experimental Nuclear Physics
1967-

Chairman

Vladimirskiy, V.V. 1967-

113 Scientific Council for "Cosmic Radiation"
? -

Chairmen

Skobel'tsyn, D.V. . . . . ? -1966
Vernov, S.N. 1966-

114 Scientific Council for the "Scientific Principles of Using Superconductivity in Energetics"
1968-

Chairman

Petrov, G.N. 1968-

115 Scientific Council for Comprehensive Power Problems
1965-

Chairman

Styrikovich, M.A. 1965-

116 Scientific Council for Neutron Physics
         1969-

Chairman

Pontekorvo, B.M. 1969-

117 Scientific Council for Applying Methods of Nuclear Physics in Adjacent Fields
         1969-

Chairman

Flerov, G.N. 1969-

118 Scientific Council for the Acceleration of Charged Particles
         1967-

Chairman

Mints, A.L. 1967-

119 Scientific Council for Problems of Electrical Metering and Data Metering Systems
         1966-

Chairmen

Shumilovskiy, N.N. 1966-1968
Sotskov, B.S. 1968-

120 Scientific Acoustics Council
         1965-1969

Chairman

Brekhovskikh, L.M. 1965-1969

121 Scientific Council for Acoustics
          1969-

          Chairman

     Rimskiy-Korsakov, A.V. 1969-

122 Scientific Council for Theoretical and Electrophysical Problems of Electrical Power Engineering
          1968-

          Chairman

     Kostenko, M.P. 1968-

123 Scientific Council for Low-Temperature Physics
          1966-

          Chairman

     Kapitsa, P.L. 1966-

124 Scientific Council for Ultrasonic Physics and Engineering
          1969-

          Chairman

     Mikhaylov, I.G. 1969-

125 Scientific Council for Nuclear Reactions
          1969-

          Chairman

     Davydov, A.S. 1969-

126 Joint Commission for the Coordination of Research on Nuclear-Gamma Resonance Spectroscopy
          1966-

Chairman

Gol'danskiy, V.I. 1966-

II. EARTH SCIENCES SECTION

* 127 Geological and Mineralogical Museum
      1917-1925

   Directors

   Vernadskiy, V.I. 1917-1921
   Borisyak, A.A. 1922-1925

* 128 Peter I Geological Museum
      1925-1930

   Director

   Levinson-Lessing, F.Yu. 1925-1930

* 129 Geological Institute
      1930-1937

   Directors

   Obruchev, V.A. 1930-1934
   Arkhangel'skiy, A.D. 1934-1937

  130 Paleozoological Institute (see No. 242)

  131 Petrographical Institute
      1930-1937

   Director

   Levinson-Lessing, F.Yu. 1930-1937

* 132 Museum of Mineralogy
      1925-1932

Director

Fersman, A.Ye. 1925-1932

* 133 M.V. Lomonosov Institute of Geochemistry, Mineralogy and Crystallography
1932-1937

Director

Fersman, A.Ye. 1932-1937

134 Institute of Geological Sciences
1937-1955

Directors

Arkhangel'skiy, A.D. 1938-1949
Varentsov, M.I. 1949-1955

* 135 Geological Institute
1955-

Directors

Varentsov, M.I. 1955-1956
Shatskiy, N.S. 1956-1961
Peyve, A.V. 1961-

* 136 Crystallography Laboratory
1917-1932
1938- ?

Director

Shubnikov, A.V. . . . . 1938- ?

* 137 Institute of Ore Deposits Geology, Petrography, Mineralogy and Geochemistry
1955-

Director

    Chukhrov, F.V. 1955-

* 138 Laboratory of Pre-Cambrian Geology
                ?   -1967

Directors

    Polkanov, A.A. 1950-1964
    Obruchev, S.V. 1964-1967

139 Institute of Pre-Cambrian Geology and Geochronology
               1967-

Director

    Krats, K.O. 1967-

* 140 Laboratory of Mineralogy and Geochemistry of Rare Elements
               ?   -1956

Director

    Vlasov, K.A. . . . . ? -1956

141 Institute of Mineralogy, Geochemistry and Crystallochemistry of Rare Elements
              1956-1961

Director

    Vlasov, K.A. 1956-1961

* 142 Geochemical Problems Laboratory
              1929-1947

Director

    Vernadskiy, V.I. 1929-1947

143 V.I. Vernadskiy Institute of Geochemistry and Analytical Chemistry
1947-

Director

Vinogradov, A.P. 1948-

* 144 Sapropelite Institute
1932-1934

Director

Zelenko, V.A. 1932-1934

* 145 Institute of Mineral Fuels
1934-1959

Directors

Gubkin, I.M. 1934-1939
Namyotkin, S.S. 1939-1948
Chernyshyov, A.B. 1948-1954
Titov, N.G. 1954-1959

* 146 Petroleum Institute
1947-1959

Directors

Namyotkin, S.S. 1948-1953
Titkov, N.I. 1953-1959

* 147 Institute of Geology and Exploitation of Mineral Fuels
1959-1961

Director

Mirchink, M.F. 1959-1961

* 148 Mining Institute
1938-1961

Directors

    Skochinskiy, A.A. 1938-1960
    Mel'nikov, N.V. 1960-1961

149 Institute of Experimental Mineralogy
            1969-

Director

    Korzhinskiy, D.S. 1969-

* 150 Hydrogeological Problems Laboratory
            ? -1961

Directors

    Slavyanov, N.N. 1947-1956
    Kamenskiy, G.N. 1956-1957
    Priklonskiy, V.A. 1957-1960
    Popov, I.V. 1960-1961

* 151 Meteorites Committee
            1939-

Chairmen

    Vernadskiy, V.I. 1939-1945
    ? 1945-1949
    Fesenkov, V.G. 1949-1964
    ? 1964-

152 Petrographic Committee
            1962-

Chairman

    Afanas'yev, V.D. 1962-

* 153 Geomorphological Institute
            1930-1934

Director

    Grigor'yev, A.A. 1930-1934

154 Institute of Sandy Deserts
        1932-1933

Director

    Komarov, V.L. 1932-1933

* 155 Institute of Physical Geography
        1934-1936

Director

    Grigor'yev, A.A. 1935-1936

156 Institute of Geography
        1937-

Directors

    Grigor'yev, A.A. 1937-1951
    Gerasimov, I.P. 1951-

* 157 Commission (from 1936 Committee) for Permafrost Studies
        1930-1939

Chairman

    Obruchev, V.A. 1930-1939

* 158 V.A. Obruchev Institute of Permafrost Studies
        1939-1961

Directors

    Obruchev, V.A. 1939-1956
    Shvetsov, G.F. 1956-1959
    Shumskiy, P.A. 1959-1961

* 159 Institute of Theoretical Geophysics
1938-1947

Director

Shmidt, O.Yu. 1938-1947

* 160 Permanent Central Seismic Commission
1900-1928

Chairmen

(Golitsyn, B.B. . . . . 1900-1916)
? 1917-1928

* 161 Seismological Institute
1928-1947

Director

Nikiforov, P.M. 1928-1947

* 162 Geophysical Institute
1947-1956

Directors

Gamburtsev, G.A. 1947-1955
? 1955-1956

163 Institute of Geophysics
1956-

Directors

Molodenskiy, M.S. 1956-1957
Karus, Ye.V. 1957-1960
Sadovskiy, M.A. 1960-

164 Institute of Atmospheric Physics
1956-

Director

    Obukhov, A.M. 1956-

* 165 Institute of Applied Geophysics
        1956-1961

Director

    Fyodorov, Ye.K. 1956-1961

* 166 Vulcanological Laboratory
        ? -1962

Directors

    Zavaritskiy, A.N. 1949-1953
    Vlodavets, V.I. 1953-1962

* 167 Marine Hydrophysical Laboratory
        ? -1948

Director

    Shuleykin, V.V. . . . . ? -1948

168 Marine Hydrophysical Institute
        1948-1961

Directors

    Shuleykin, V.V. 1948-1956
    Grabovskiy, V.I. 1956-1961

* 169 Oceanological Laboratory
        1941-1946

Directors

    ? 1941-1946

170 Institute of Oceanology
           1946-

Directors

    ? 1946-1949
    Shirshov, P.P. 1949-1953
    Kort, V.G. 1953-1965
    Monin, A.S. 1965-

* 171 Laboratory of Lake Studies
             ?   -

Directors

    Nalivkin, D.V. 1946-1955
    Kalesnik, S.V. 1955-   ?
    ? 1969-

172 Institute of Water Problems
            1967-

Director

    Voznesenskiy, A.N. 1968-

* 173 Baykal Limnological Station
             ?   -1961

Directors

    Vereshchagin, G.Yu. 1929-1944
    ? 1944-1961

* 174 V.V. Dokuchayev Soil Institute
             1925-1961

Directors

    Levinson-Lessing, F.Yu. 1925-1927
    Glinka, K.D. 1927
    Gedroyts, K.K. 1928-1930
    Keller, B.A. 1931-1936
    Prasolov, L.I. 1937-1949
    Tyurin, I.V. 1949-1961

* 175 V.V. Dokuchayev Central Museum of Soil Science, Leningrad
                1918-

Directors

```
?              . . . . . . . . . . . . . . . 1918-1947
Shokal'skaya, Z.Yu. . . . . . . . . . . 1947-1959
?              . . . . . . . . . . . . . . . 1959-
```

* 176 Tectonics Commission
                1948-1961

Chairman

Obruchev, V.A. 1948-   ?

   177 Commission for International Tectonic Charts
                1961-

Chairmen

?   . . . . ? - ?

* 178 Committee for Geodesy and Geophysics
                1955-

Chairman

Belousov, V.V. 1955-   ?

* 179 Antarctic Research Council
                1955-1959

Chairman

Shcherbakov, D.I. 1955-1959

   180 Interdepartmental Commission for Antarctic Research
                1959-

Chairmen

> ? 1959-1966
> Avsyuk, G.A. 1966-

181 Commission for Problems of Natural Waters Protection
   1969-

Chairman

Semyonov, N.N. 1969-

* 182 Scientific Council for Soil Science and Melioration Problems
   1968-

Chairman

Kovda, V.A. 1968-

183 Comprehensive Antarctic Expedition
   1955-

Director

Somov, M.M. 1955-

III. CHEMOTECHNOLOGICAL AND BIOLOGICAL SCIENCES SECTION

184 Leningrad Institute of High-Molecular Compounds
   1946-

Directors

Ushakov, S.N. 1948-1953
Danilov, S.N. 1953-1960
Koton, M.M. 1960-

185 A.V. Topchiyev (since 1963) Institute of Petrochemical Synthesis
1959-

Directors

Topchiyev, A.V. 1959-1962
Namyotkin, N.S. 1962-

* 186 N.D. Zelinskiy Institute of Organic Chemistry
1934-

Directors

Favorskiy, A.Ye. 1934-1939
Nesmeyanov, A.N. 1939-1954
Kazanskiy, B.A. 1954-1966
Kochetkov, N.K. 1966-

* 187 Laboratory for the Study and Synthesis of Vegetable and Animal Products
? -1934

Directors

? . . . . ? -1934

* 188 A.Ye. Arbuzov Chemical Institute, Kazan'
1962-1965

Director

Arbuzov, A.Ye. 1962-1965

189 Institute of Organic Chemistry, Kazan'
? -1965

Director

Arbuzov, B.A. 1959-1965

190 A.Ye. Arbuzov Institute of Organic and Physical Chemistry, Kazan'
1965-

Director

Arbuzov, B.A. 1965-

* 191 Laboratory of Colloidal Electrochemistry
1930-1934

Director

Kistyakovskiy, V.A. 1930-1934

* 192 Institute of Colloidal Electrochemistry
1934-1945

Directors

Kistyakovskiy, V.A. 1934-1939
Frumkin, A.N. 1939-1945

193 Institute of Physical Chemistry
1945-

Directors

Frumkin, A.N. 1945-1949
Akimov, G.V. 1949-1954
Spitsyn, V.I. 1954-

* 194 Institute of Chemical Physics
1931-

Director

Semyonov, N.N. 1931-

* 195 Laboratory of Anisotropic Structures
?   -1959

Directors

Burov, A.K. 1953-1958
Semyonov, N.N. 1958-1959

196 Institute of Electrochemistry
        1957-

Director

Frumkin, A.N. 1958-

197 Institute of Elementary Organic Compounds
        1954-

Director

Nesmeyanov, A.N. 1954-

* 198 I.I. Baykov Metallurgical Institute
        1939-

Directors

Bardin, I.P. . . . . 1939- ?
Pridantsev, M.V. . . . .   ? -1968
Samarin, A.M. 1968-1970
Ageyev, N.V. 1970-

* 199 Magnetics Laboratory
        1957-1961

Directors

Kondorskiy, Ye.I. 1957-1960
Panchenko, V.D. 1960-1961

200 Institute of New Chemical Problems
        1964-

Director

Zhavoronkov, N.M. 1964-

* 201 Laboratory of General Chemistry (former Central Chemical Laboratory)
1918-1934

Director

Kurnakov, N.S. 1920-   ?

* 202 Institute of Physicochemical Analysis
1918-1934

Directors

?   1918-1934

203 Institute for the Study of Platinum and Other Precious Metals
1918-1934

Directors

Chugayev, L.A. 1918-1922
Kurnakov, N.S. 1922-1934

204 N.S. Kurnakov (since 1944) Institute of General and Inorganic Chemistry
1934-

Directors

Kurnakov, N.S. 1934-1941
Chernyayev, I.I. 1941-1960
Zhavoronkov, N.M. 1960-

* 205 I.V. Grebenshchikov (since 1962) Institute of Silicate Chemistry
1948-

Directors

Grebenshchikov, I.V. 1948-1957
Toropov, N.A. 1957-1968
Keller, E.K. 1968-

* 206　Laboratory of Polymer Stabilization, Gorky
　　　　　　　1963-1969

　　　　　　Director

　　　　Razuvayev, G.A. 1963-1969

　　207　Institute of Chemistry, Gorky
　　　　　　　1969-

　　　　　　Director

　　　　Razuvayev, G.A. 1969-

* 208　Hydrochemical Institute, Novocherkassk
　　　　　　　1938-1961

　　　　　　Directors

　　　　Kashinskiy, P.A. 1938-1951
　　　　Alekin, O.A. 1951-1960
　　　　Datsko, G.V. 1960-1961

　　209　Commission for the Automation of Chemical Processes
　　　　　　　1962-

　　　　　　Chairman

　　　　Zhavoronkov, N.M. 1962-

* 210　Institute of Biological Physics, Pushchino
　　　　　　　1952-

　　　　　　Directors

　　　　Kuzin, A.M. 1952-1958
　　　　Frank, G.M. 1958-

* 211　Laboratory of Animal Biochemistry and
　　　　Physiology
　　　　　　　?　-1934

Director

  Gulevich, V.S. 1933-1934

* 212 Laboratory of Animal Physiology
    1934-1936

Director

  Orbeli, L.A. 1934-1936

* 213 A.N. Bakh Institute of Biochemistry
    1934-

Directors

  Bakh, A.N. 1935-1946
  Oparin, A.I. 1946-

 214 Institute of Biochemistry and Physiology of Microorganisms
    1965-

Directors

  Iyerusalimskiy, N.D. 1965-1968
  Skryabin, G.K. 1968-

* 215 Microbiological Laboratory
    1929-1934

Director

  Nadson, G.A. 1929-1934

 216 Institute of Microbiology
    1934-

Directors

  Nadson, G.A. 1934-1938
  Isachenko, B.L. 1938-1949
  Imshenetskiy, A.A. 1949-

* 217 Institute of Radiation and Physicochemical Biology
    1957-1965

    Director

  Engel'gardt, V.A. 1957-1965

218 Institute of Molecular Biology
    1965-

    Director

  Engel'gardt, V.A. 1965-

* 219 Laboratory of Electronic Microscopy
    1948-1967

    Directors

  ? 1948-1967

* 220 Laboratory of Plant Biochemistry and Physiology
    1922-1934

    Directors

  Famintsyn, A.S. 1917-1919
  Borodin, I.P. 1919-1920
  Palladin 1920-1922
  Kostychev, S.P. 1922-1932
  Rikhter, A.A. 1932-1934

* 221 K.A. Timiryazev Institute of Plant Physiology
    1934-

    Directors

  Rikhter, A.A. 1934-1938
  Bakh, A.N. 1938-1946
  Maksimov, N.A. 1946-1952
  Kursanov, A.L. 1952-

222 Institute of Natural Compounds Chemistry
        1959-

Director

Shemyakin, M.M. 1959-1970
? 1970-

* 223 Protein Institute
        1967-

Director

Spirin, A.S. 1967-

224 Institute of Photosynthesis
        1966-

Director

Yevstigneyev, V.B. 1966-

* 225 Physiological Laboratory
        1889-1925

Director

Pavlov, I.P. 1917-1925

* 226 I.P. Pavlov Physiological Institute
        1925-1934
        1936-1950

Directors

Pavlov, I.P. 1925-1934
Orbeli, L.A. 1936-1950

227 Institute of Physiology and Pathology of Higher
     Nervous Activity
        1934-1936

Director

    Pavlov, I.P. 1934-1936

* 228   I.P. Pavlov Institute of Physiology
          1950-

Directors

    Bykov, K.M. 1950-1959
    Chernigovskiy, V.N. 1959-

* 229   Institute of Higher Nervous Activity
          1950-1960

Directors

    Asratyan, E.A. 1950-1952
    Ivanov-Smolenskiy, A.G. 1952-1957
    Rusinov, V.S. 1957-1958
    Voronin, L.G. 1958-1960

230   Physiological Laboratory
          ?   -1960

Director

    Asratyan, E.A. 1944-1960

231   Institute of Higher Nervous Activity and Neurophysiology
          1960-

Director

    Asratyan, E.A. 1960-

* 232   Moscow Physiological Institute
          ?   -1948

Director

    Shtern, L.S. 1940-1948

* 233 Laboratory for the Study of Nervous and Humoral Regulation
1963-1969

Directors

Grashchenkov, N.I. 1963-1969
Parin, V.V. 1969

234 N. I. Grashchenkov Laboratory of Control Problems in Human and Animal Organisms
1969-

Director

Parin, V.V. 1969-

* 235 Laboratory of Evolutionary Physiology
1954-1956

Directors

? 1954-1956

* 236 I. M. Sechenov Institute of Evolutionary Physiology
1956-1964

Directors

Orbeli, L.A. 1956-1958
Ginetsinskiy, A.G. . . . . ? -1960
Kreps, Ye.M. 1960-1964

237 Institute of Evolutionary Physiology and Biochemistry
1964-

Director

Kreps, Ye.M. 1964-

* 238 N.A. Morozov "Borok" Biological Station
1947-1956

Directors

Shennikov, A.P. 1947-1953
Papanin, I.D. 1953-1956

* 239 Institute of Water Reservoirs Biology
1956-1962

Director

Papanin, I.D. 1956-1962

240 Institute of Inland Waters Biology
1962-

Director

Papanin, I.D. 1962-

* 241 Laboratory of Evolutionary Morphology
1930-1934

Director

Severtsov, A.N. 1930-1934

242 Paleozoological Institute
1930-1934

Directors

? . . . . ?

* 243 Institute of Evolutionary Morphology and Paleozoology
1934-1937

Director

Severtsov, A.N. 1934-1937

244 Institute of Paleontology
          1937-

     Directors

     Borisyak, A.A. 1937-1944
     Orlov, Yu.A. 1945-1969
     Kramarenko, N.N. 1969-

* 245 A.N. Severtsov Institute of Evolutionary Morphology
          1937-1949

     Director

     Shmal'gauzen, I.I. 1937-1948

* 246 Institute of Experimental Biology
          1938-1940

     Directors

     ? . . . . ?

* 247 Institute of Cytology, Histology and Embryology
          1940-1949

     Director

     Khrushchov, G.K. 1940-1949

* 248 A.N. Severtsov Institute of Animal Morphology
          1949-1968

     Director

     Khrushchov, G.K. 1949-1967

249 A.N. Severtsov Institute of Evolutionary Morphology and Ecology of Animals
          1968-

Director

    Sokolov, V.Ye. 1968-

250 N. K. Kol'tsov Institute of Biological Development
        1968-

Director

    Astaurov, B.L. 1968-

251 Institute of Cytology
        1957-

Directors

    Nasonov, D.N. 1957-1959
    Troshin, A.S. 1959-

* 252 Botanical Museum
        1824-1931

Director

    Litvinov, D.I. 1917-1929

* 253 Main Botanical Gardens, Leningrad
        1929-1931

Director

    Isachenko, B.L. 1917-1930

254 V.L. Komarov Botanical Institute
        1931-

Directors

    Keller, B.A. 1931-1937
    Shishkin, B.K. 1938-1949
    Kuprevich, V.F. 1949-1952
    Baranov, P.A. 1952-1963
    Fyodorov, A.A. 1963-

255 Main Botanical Gardens, Moscow
1945-

Director

Tsitsin, N.V. 1945-

* 256 Forest Institute
1944-1959

Director

Sukachyov, V.N. 1944-1959

* 257 Forestry Laboratory
1959-1970

Directors

Sukachyov, V.N. 1959-1966
Molchanov, A.A. 1966-1970

258 Institute of Forestry
1970-

Director

Molchanov, A.A. 1970-

* 259 Institute of Timber and Timber Chemistry, Arkhangel'sk
1958-   ?

Director

Melekhov, I.S. 1958-   ?

* 260 Genetics Laboratory
1930-1933

Director

Vavilov, N.I. 1930-1933

* 261 Institute of Genetics
1934-1966

Directors

Vavilov, N.I. 1934-1940
Lysenko, T.D. 1940-1965
Dubinin, N.P. 1965-1966

262 Institute of General Genetics
1966-

Director

Dubinin, N.P. 1966-

* 263 Zoological Museum
1832-1931

Directors

Nasonov, N.V. 1917-1921
Byalynitskiy-Birulya, A.A. 1921-1929

264 Zoological Institute, Leningrad
1931-

Directors

Zernov, S.A. 1931-1942
Pavlovskiy, V.N. 1942-1962
Bykhovskiy, B.Ye. 1962-

* 265 A.O.Kovalevskiy Biological Station, Sevastopol'
1871-1938
1944-1957

Directors

Nikitin, V.N. 1922-1925
Nasonov, N.V. 1925-1930
Zernov, S.A. 1930-1938
Vodyanitskiy, V.A. 1944-1957

266 Institute of South Sea Biology
1957-   ?

Director

Vodyanitskiy, V.A. 1957-   ?

267 Helminthological Laboratory
?   -

Director

Skryabin, K.I. 1942-

* 268 Murmansk Biological Station
?   -1938
1949-1953

Directors

?   . . . . ?   -1938
Kuznetsov, V.V. 1949-1953

* 269 Scientific Directors' Council of the Biological Research Center, Pushchino
1962-

Chairmen

Frank, G.M. . . . . ?   -1968
Skryabin, G.K. 1968-

270 Scientific Council for Pure Substances and Physicochemical Methods of Analysis
?   -

Chairmen

Vinogradov, A.P. . . . . ?   -1967
Alimarin, I.P. 1967-

271 Scientific Council for the Physicochemical Principles of Metallurgical Processes
? -

Chairman

Samarin, A.M. 1966-1970

272 Scientific Council for the Physical Chemistry of Ionic Fusions and Solid Electrolytes
1969-

Chairman

Delimarskiy, Yu.K. 1969-

273 Scientific Council for the Chemistry of Solid Mineral Fuels
1964-

Chairman

Karavayev, N.M. 1964-

274 Scientific Council for Chemical Thermodynamics and Thermochemistry
1967-

Chairman

Gerasimov, Ya.I. 1967-

275 Scientific Council for Organic Elemental Chemistry
? -

Chairmen

Nesmeyanov, A.N. . . . . ? -1965
Kabachnik, M.I. 1965-

276 Scientific Council for Petrochemistry
? -

Chairmen

       Lavrovskiy, K.P. . . . . ? -1966
       Kamzolkin, V.V. 1966-1969
       Namyotkin, N.S. 1969-

277 Scientific Council for Genetics and Selection Problems
              ? -

Chairmen

       Dubinin, N.P. 1965-1968
       Belyayev, D.K. 1968-

278 Scientific Council for the Microbiological Synthesis of Protein and Other Products from Hydrocarbons
              1965-

Chairmen

       ? 1965-1968
       Skryabin, G.K. 1968-

279 Scientific Council for Theoretical Problems of Biological Damage of Materials
              1967-

Chairman

       Flerov, B.K. 1967-

280 Joint Scientific Council for Human and Animal Physiology
              ? -

Chairmen

       ? . . . . ? - ?

281 Scientific Council for Comprehensive Biogeocenological Research of Living Nature and Scientific Principles for Mastering and Protecting It
                ? -

Chairman

Lavrenko, Ye.M. 1968-

IV. *SOCIAL SCIENCES SECTION*

* 282 Asiatic Museum
            1818-1930

Directors

? . . . . ? - ?

* 283 Institute of Buddhist Culture
            ? -1930

Directors

? . . . . ? - ?

* 284 Turkological Study Center
            ? -1930

Directors

? . . . . ? - ?

* 285 Pacific Institute
            ? -1950

Director

Zhukov, Ye.M. 1943-1950

* 286   Institute of Oriental Studies
                1930-1960
                1968-

         Directors

         Ol'denburg, S.F. 1930-1932
         Samoylovich, A.N. 1932-1937
         Barannikov, A.P. 1938-1941
         Struve, V.V. 1941-1950
         Tolstov, S.P. 1950-1952
         Avdiyev, V.I. 1952-1954
         Guber, A.A. 1955-1956
         Gafurov, B.G. 1956-1960
         Gafurov, B.G. 1968-

  287   Sinological Institute
                1956-1960

         Directors

         Perevertaylo, A.S. 1956-1960
         Tikhvinskiy, S.L. 1960

  288   Institute of Asian Peoples
                1960-1968

         Director

         Gafurov, B.G. 1960-1968

* 289   African Institute
                1959-

         Directors

         Potekhin, I.I. 1959-1964
         Solodovnikov, V.G. 1964-

* 290   Scientific Council for African Problems
                1966-

Chairmen

Rumyantsev, A.M. 1966
Solodovnikov, V.G. 1966-

291 Institute of Latin America
         1961-

Directors

Mikhaylov, S.S. 1961-1965
Rudenko, B.T. 1965-1966
Vol'skiy, V.V. 1966-

292 Institute of the United States of America
         1967-

Director

Arbatov, G.A. 1967-

* 293 Anthropological and Ethnographical Museum
         1878-1933

Directors

? . . . . ? - ?

* 294 Commission for the Study of the Tribal Structure of the Population of Russia and Adjoining Countries
         1917-1930

Chairmen

? . . . . ? - ?

295 Institute for the Study of the Peoples of the USSR
         1930-1933

Directors

          ?   . . . .   ? - ?

* 296   Institute of Anthropology, Archeology and Ethnography
                    1930-1937

            Directors

            ?   1930-1933
            Motorin, N.M. 1933-1934
            Meshchaninov, I.I. 1934-1937
            Struve, V.V. 1937

* 297   Museum of the History of Religion
                    1932-1936

            Director

            Tan-Bogoroz, V.G. 1932-1936

* 298   Miklukho-Maklay Institute of Ethnography
                    1937-

            Directors

            Struve, V.V. 1937-1940
            Vinnikov, I.N. 1941
            Abramzon, S.M. 1941-1942
            Tolstov, S.P. 1942-1966
            Bromley, Yu.V. 1966-

* 299   Institute of the History of Material Culture
                    1937-1959

            Directors

            Orbeli, I.A. 1937-1939
            Artamonov, M.I. 1939-1943
            Grekov, B.D. 1943-1946
            Udal'tsov, A.D. 1946-1956
            Rybakov, B.A. 1956-1959

300 Institute of Archeology
          1959-

   Director

   Rybakov, B.A. 1959-1968

* 301 History of Knowledge Commission
          1926-1932

   Chairmen

   Vernadskiy, V.I. 1926-1930
   ?   1930-1932

* 302 Institute of the History of Science and Engineering
          1932-1938

   Directors

   ?   1932-1936
   Osinskiy, V.V. 1937-1938

* 303 Commission for the History of Engineering
          1938-1953

   Chairmen

   ?   1938-1953

* 304 Institute of Natural Sciences History
          1944-1953

   Director

   Koshtoyants, Kh.S. 1946-1953

* 305 Institute of the History of Natural Science and Engineering
          1953-

Directors

Samarin, A.M. 1953-1955
Kuznetsov, I.V. 1955-1956
Figurovskiy, N.A. 1956-1962
Kedrov, B.M. 1962-

306 Commission for the History of Geological and Geographical Sciences
1948-1953

Chairman

Zavaritskiy, A.N. 1948-1953

307 Commission for the History of Chemical Sciences
? -1953

Chairmen

? . . . . ? - ?

308 Commission for Studying and Publishing A.M.Butlerov's Scientific Legacy
? -1953

Chairmen

? . . . . ? - ?

309 Commission for Studying and Publishing M.V. Lomonosov's Scientific Legacy
? -1953

Chairmen

? . . . . ? - ?

310 Commission for Studying and Publishing D.I.Mendeleyev's Scientific Legacy
? -1953

Chairmen

    ? . . . . ? - ?

311 M.V. Lomonosov Museum
        1947-1953

Directors

    ? . . . . ? - ?

312 Commission for the History of the USSR Academy of Sciences
        1941-1953

Chairmen

Vavilov, S.I. 1941-1953
Volgin, V.P. 1953

313 Institute of Art History
        1944-1961

Director

Grabar', I.E. 1944-1961

* 314 Archeographic Commission
        1922-1926
        1956-

Chairmen

? 1922-1926
? 1956-1966
Shunkov, V.I. 1966-1969
Shmidt, S.O. 1969-

315 Archeographic History Commission
        1926-1931

Chairmen

? . . . . ? - ?

* 316 Institute of the History of Archeography
1931-1936

Directors

? 1931-1933
Tomsinskiy, S.G. 1934
Volgin, V.P. 1935-1936

317 Historical Commission
1933-1936

Chairman

Volgin, V.P. 1933-1936

* 318 Institute of History
1936-1968

Directors

Lukin, N.M. 1936-1937
Grekov, B.D. 1937-1953
Sidorov, A.L. 1953-1959
Khvostov, V.M. 1959-1967

319 Institute of the History of the USSR
1968-

Directors

Rybakov, B.A. 1968-1970
Volobuyev, P.V. 1970-

320 Institute of Universal History
1968-

Director

    Zhukov, Ye.M. 1968-

* 321 Institute of Slavic Studies
        1947-1969

Directors

    Grekov, B.D. 1947-1951
    Tret'yakov, P.N. 1951-1959
    Udal'tsov, I.I. 1959-1962
    Khrenov, I.A. 1962-1969

  322 Institute of Slavic and Balkan Studies
        1969-

Director

    Markov, D.F. 1969-

* 323 Institute of Philosophy
        1936-

Directors

    Adoratskiy, V.V. 1936-1939
    Yudin, P.F. 1939-1944
    Svetlov, V.I. 1944-1946
    Aleksandrov, G.F. 1947-1954
    Fedoseyev, P.N. 1955-1962
    Konstantinov, F.V. 1962-1968
    Kopnin, P.V. 1968-

* 324 Institute of Specific Social Research
        1968-

Director

    Rumyantsev, A.M. 1968-

* 325 Institute of State Law
        1936-1938

Director

Vyshinskiy, A.Ya. 1936-1938

* 326   A.Ya. Vyshinskiy Institute of Law
        1938-1960

Directors

Vyshinskiy, A.Ya. 1938-1941
Traynin, I.P. 1942-1947
Korovin, Ye.A. 1948-1952
Orlovskiy, P.Ye. 1952-1958
Romashkin, P.S. 1958-1960

  327   Institute of State and Law
        1960-

Directors

Romashkin, P.S. 1960-1964
Chkhikvadze, V.M. 1964-

* 328   Institute of World Economics and World Politics
        1936-1947

Director

Varga, Ye.S. 1936-1947

* 329   Institute of Economics
        1936-

Directors

Savel'yev, M.A. 1936-1938
Milyutin, N.A. 1938-1940
Markus, B.L. 1940-   ?
Khromov, P.A. 1947
Ostrovityanov, K.V. 1947-1953
D'yachenko, V.P. 1954-1956
Laptev, I.D. 1956-1959
Plotnikov, B.N. 1959-1965
Gatovskiy, L.M. 1965-

330 Institute of World Economics and International Relations
1956-

Directors

Arzumanyan, A.A. 1956-1966
Inozemtsev, N.N. 1966-

331 Institute of Economics of the World Socialist System
1960-

Directors

Sorokin, G.M. 1961-1969
Bogomolov, O.T. 1969-

332 Laboratory of Economic Mathematical Methods
? -1963

Director

Nemchinov, V.S. 1962-1963

* 333 Central Economic Mathematics Institute
1963-

Director

Fedorenko, N.P. 1963-

Leningrad Branch of the Central Economic Mathematics Institute
1965-

334 Institute of Economics, Ufa
1965-

Directors

? . . . . ? - ?

335 Institute of the Far East
1968-

Director

Sladkovskiy, M.I. 1968-

* 336 A.M. Gorky Institute of World Literature
1938-

Directors

Luppol, I.K. 1938-1946
Shishmarev, V.F. 1947-1948
Yegolin, A.M. 1948-1952
Anisimov, I.I. 1952-1966
   ? 1966-1968
Suchkov, B.L. 1968-

* 337 Pushkin House
1907-1930

Directors

Kotlyarevskiy, N.A. 1917-1925
   ? 1925-1930

* 338 Institute of New Russian Literature
1930-1932

Director

Lunacharskiy, A.V. 1930-1932

339 Institute of Russian Literature
1932-

Directors

Lunacharskiy, A.V. 1932-1933
   ? 1933-1934
Gorky, A.M. 1935-1936
Lebedev-Polyanskiy, P.I. 1937-1948
Bel'chikov, N.F. 1949-1955

　　　　Bushmin, A.S. 1955-1965
　　　　Bazanov, V.G. 1965-

* 340　Caucasian Institute of the History of Archeology
　　　　　　　　1917-1921

　　　　　　　　　Director

　　　　Marr, N.Ya. 1917-1921

　341　Institute of Japhetic Studies
　　　　　　　　1921-1922

　　　　　　　　　Director

　　　　Marr, N.Ya. 1921-1922

　342　Japhetic Institute
　　　　　　　　1922-1931

　　　　　　　　　Director

　　　　Marr, N.Ya. 1922-1931

* 343　N.Ya. Marr Institute of Language and Thinking
　　　　　　　　1931-1950

　　　　　　　　　Directors

　　　　Marr, N.Ya. 1931-1934
　　　　Meshchaninov, I.I. 1934-1950

* 344　Institute of Russian Language
　　　　　　　　?　　-1950
　　　　　　　　1958-

　　　　　　　　　Directors

　　　　Obnorskiy, S.P. 1944-1950
　　　　Vinogradov, V.V. 1950-1968
　　　　Trubachyov, O.N. 1968
　　　　Filin, F.P. 1969-

345 Institute of Linguistics
            1950-

   Directors

   Vinogradov, V.V. 1950-1954
   Borkovskiy, V.I. 1954-1960
   Serebrennikov, B.A. 1960-1964
   Filin, F.P. 1964-1969

* 346 Institute of Speech and Written Language
            1938-   ?

   Director

   Petrosyan, V.A. 1938-   ?

* 347 Galimdzhan Ibragimov (since 1969) Institute of Language, Literature and History, Kazan'
            1963-

   Director

   Mukharyamov, M.K. 1964-

* 348 Petrozavodsk Institute of Language, Literature and History
            1963-1967

   Directors

   Mashezerskiy, V.I. 1963-1965
   Vlasova, M.N. 1965-1967

* 349 Ufa Institute of History, Language and Literature
            1963-1967

   Director

   Sayranov, Kh.S. 1964-1967

350 Commission for Regulating the Spelling and Pronunciation of Foreign Proper and Geographical Names
   1960-

   Chairman

   Barkhudarov, S.G. 1960-

351 Scientific Economic Commission
   1966-

   Chairman

   Gatovskiy, L.M. 1966-

* 352 Commission for the History of World Culture
   1959-   ?

   Chairmen

   ?   . . . . ?   -   ?

353 Scientific Council for the History of World Culture
   1969-

   Chairman

   Konrad, N.I. 1969-

354 Scientific Council for Dialectology and Language History
   1965-

   Chairman

   Avanesov, R.I. 1965-

355 Scientific Council for Information in the Field of Social Sciences
   1965-

Chairmen

    Shunkov, V.I. 1965-1967
      ?   1967-

356 Scientific Council for the History of Social Thought
    1966-

Chairman

    Iovchuk, M.T. 1966

357 Scientific Council for Historical Geography and Cartography
    1968-

Chairman

    Beskrovnyy, L.G. 1968-

358 Scientific Council for Laws Governing the Development of Social Relations and Spiritual Life of Socialist Society
    1966-

Chairman

    Stepanyan, Ts.A. 1967-

359 Scientific Council for the World Socialist System
    1967-

Chairman

    Sorokin, G.M. 1967-

360 Scientific Council for Nationality Problems
    1969-

Chairman

Zhukov, Ye.M. 1969-

361 Scientific Council for Problems of Foreign Ideological Trends
1966-

Chairman

Mitin, M.B. 1966-

362 Scientific Council for Aesthetic Problems
1967-

Chairman

Yegorov, A.G. 1967-

363 Scientific Council for Laws Governing the Development of State, Administration and Law
1966-

Chairman

Chkhikvadze, V.M. 1966-

364 Scientific Folklore Council
? -

Chairmen

Bazanov, V.G. . . . . ?   -1967
Kravtsov, N.I. 1967-

365 Scientific Council for the Social and Economic Problems of Population
1967-

Chairman

Ryabushkin, T.V. 1967-

366 Council for Organizing the Acquisition and Use of Archive Material
1964- ?

Chairman

Agoshkov, M.I. . . . . 1964- ?

367 Scientific Council for the "Optimum Planning and Administration of the National Economy"
1968-

Chairman

Fedorenko, N.P. 1968-

368 Institute of International Workers' Movement
1968-

Director

Timofeyev, T.T. 1968-

* 369 Archives
1828-

Directors

? 1917-1933
Knyazev, G.A. 1933-1963
Levshin, B.V. . . . . 1963- ?

* 370 Main Library of the Social Sciences Department
1918-

Directors

Ivanov, D.D. 1921-1946
? 1946-1970
Delyusin, L.P. 1970-

Branches and Bases of the USSR Academy of Sciences

* 371   V.L. Komarov Far Eastern Branch
                1932-1957

      Presidium Chairmen

    Komarov, V.L. 1932-1945
    ?   1945-1957

* 372   Transcaucasian Branch
                1932-1934

      Presidium Chairman

    Marr, N.Ya. 1932-1934

* 373   Azerbaydzhani Branch
                1935-1945

      Presidium Chairmen

    Levinson-Lessing, F.Yu. 1935-1937
    Gubkin, I.M. 1937-1940
    Namyotkin, S.S. 1940-1945

* 374   Armenian Branch
                1935-1943

      Presidium Chairmen

    Levinson-Lessing, F.Yu. . . . . 1935- ?
    Fersman, A.Ye. . . . .   ? -1937
    Orbeli, I.A. 1938-1943

* 375   Georgian Branch
                1935-1941

      Presidium Chairmen

    Gorbunov, N.P. 1935-1937
    Muskhelishvili, N.I. 1938-1941

* 376 Ural Branch
            1932-

   Presidium Chairmen

   Fersman, A.Ye. 1932-1938
   Bardin, I.P. 1938-1957
   Demenev, N.V. 1958-1965
   Spasskiy, S.S. 1965-

* 377 Kazakhstani Research Base
            1932-1935

   Director

   Samoylovich, A.N. 1932-1935

* 378 Kazakh Branch
            1935-1946

   Presidium Chairmen

   Arkhangel'skiy, A.D. 1935-1940
       ?   1940-1946

* 379 Tadzhikistani Research Base
            1932-1941

   Directors

   Rikhter, A.A. 1932-1935
   Gorbunov, N.P. 1936-1938
   Pavlovskiy, Ye.N. 1938-1941

* 380 Tadzhik Branch
            1941-1951

   Presidium Chairman

   Pavlovskiy, Ye.N. 1941-1951

* 381 S.M. Kirov Kola Research Base
            1934-1950

Directors

Fersman, A.Ye. 1934-1941
Belyankin, D.S. . . . . ? -1950

382 Kola Branch
1950-

Presidium Chairmen

Belyankin, D.S. 1950-1952
Sidorenko, A.V. 1952-1964
Kozlov, Ye.K. . . . . 1964- ?

383 Northern Research Base
1936-1944

Directors

Knipovich, N.M. 1936-1939
? 1939-1944

* 384 Komi Research Base
1944-1949

Directors

? 1944-1949

385 Komi Branch
1949-

Presidium Chairmen

? 1949-1953
Sirin, N.A. 1953-1956
Vavilov, P.P. 1956-1965
Podoplelov, V.P. . . . . 1965- ?

* 386 Uzbek Branch
1940-1943

Presidium Chairmen

    ? 1939-1943

* 387 Turkmen Branch
        1941-1951

Presidium Chairmen

Keller, B.A. 1941-1945
Nalivkin, A.V. 1946-1951

* 388 Kirghiz Branch
        1943-1954

Presidium Chairmen

Skryabin, K.I. 1943-1952
Akhunbayev, I.K. 1952-1954

* 389 West Siberian Branch
        1944-1957

Presidium Chairmen

Skochinskiy, A.A. 1944-1951
Shmargunov, K.N. 1951-1953
Gorbachyov, T.F. 1954-1957

* 390 Kazan' Branch
        1945-1963

Presidium Chairman

Arbuzov, A.Ye. 1945-1963

* 391 Bashkir Branch
        1951-1963
        1967-

Presidium Chairmen

Vakhrushev, G.V. 1951-1957
Obolentsev, R.D. 1957-1963
    ? 1967-

* 392 East Siberian Branch
                1949-1957

          Presidium Chairmen

          Zvonkov, V.V. 1950-1954
          Titov, N.G. 1954-1955
          Pavlovskiy, Ye.V. 1955-1956
          Krotov, V.A. 1956-1957

* 393 Daghestani Research Base
                1946-1950

          Directors

          ?   1946-1950

  394 Daghestani Branch
                1950-

          Presidium Chairmen

          Meshchaninov, I.I. 1950-1953
          Amirkhanov, Kh.I. 1953-   ?

* 395 Karelo-Finnish Research Base
                1946-1949

          Directors

          ?   1946-1949

* 396 Karelo-Finnish Branch
                1949-1956

          Presidium Chairmen

          Gorskiy, I.I. 1949-1952
          Syukiyaynen, I.I. 1952-1956

* 397 Karelian Branch
                1956-1963
                1967-

Presidium Chairmen

    Syukiyaynen, I.I. 1956-1957
    Slodkevich, V.S. 1957-1960
    Dadykin, V.P. 1960-1963
    P'yavchenko, N.I. 1967-

* 398 Yakutian Research Base
             1947-1949

Director

    Tsytovich, N.A. 1947-1949

  399 Yakutian Branch
             1949-1957

Presidium Chairmen

    Tsytovich, N.A. 1949-1953
    Dadykin, V.P. 1953-1957

* 400 Sakhalin Research Base
             1946-1949

Director

    Mironov, S.I. 1946-1949

  401 Sakhalin Branch
             1949-1955

Presidium Chairmen

    Mironov, S.I. 1949-1951
    Klimov, B.K. 1951-1954
    Goremykin, V.I. 1954-1955

* 402 Moldavian Research Base
             1946-1949

Directors

    ? 1946-1949

* 403 Moldavian Branch
          1949-1960

Presidium Chairmen

    Baranov, P.A. 1949-1955
    Grosul, Ya.S. 1955-1960

* 404 Crimean Research Base
          1947-1949

Director

    Kozin, Ya.D. 1947-1949

* 405 Crimean Branch
          1949-1954

Presidium Chairmen

    Udal'tsov, A.D. 1949-1953
    Pavlovskiy, Ye.N. 1953-1954

Siberian Division of the USSR Academy of Sciences

1957-

On 18 May 1957 the USSR Council of Ministers passed a resolution on the organization of the Siberian Division of the USSR Academy of Sciences and the construction of an academic township near Novosibirsk. The Siberian Division included the former West Siberian, East Siberian, Yakutian and Far Eastern Branches of the USSR Academy of Sciences.

The Siberian Division differs from the other specialized Departments of the Academy and may be regarded as constituting the RSFSR Academy of Sciences.

Unlike the other Departments of the Academy, the Siberian Division is subordinate not only to the USSR Academy of Sciences' Presidium, but also to the RSFSR Council of Ministers. The executive organ of the Siberian Division is its Presidium, headed by the Division's Chairman, Academician M. A. Lavrent'yev, who is simultaneously a vice-president of the USSR Academy of Sciences.

PRESIDIUM OF THE USSR ACADEMY OF SCIENCES' SIBERIAN DIVISION

Chairman

Lavrent'yev, M.A. 1957-

Deputy Chairmen

Khristianovich, S.A. 1958-
Trofimuk, A.A. 1958-
Gorbachyov, T.F. 1958-
Novikov, I.I. 1958-

Presidium Members

Bogolyubov, N.N. 1958-1963
Vekua, I.I. 1958-
Sobolev, S.L. 1958-
Kuznetsov, V.D. 1958-1963
Budker, G.I. 1958-
Dubinin, N.P. 1958-1966
Koval'skiy, A.A. 1958-
Nekrasov, N.N. 1958-
Nikolayev, A.V. 1958-
Piyp, B.I. 1958-1966
Zhukov, A.B. 1959-
Prudenskiy, G.A. 1958-1967
Khel'kvist, G.A. 1958-1968
Kirenskiy, L.V. 1958-1969
Bykov, V.T. 1958-
Odintsov, M.M. 1958-

Rozhkov, I.S. 1958-1965
Melent'yev, L.A. 1960-
Khomentovskiy, A.S. 1960-

*JOINT LEARNED COUNCILS OF THE USSR ACADEMY
OF SCIENCES' SIBERIAN DIVISION*

406 Joint Learned Council for Physical, Mathematical and Technical Sciences
     1958-

     Chairman

     Sobolev, S.L. 1958-

407 Joint Learned Council for Chemical Sciences
     1958-

     Chairman

     Nikolayev, A.V. 1958-

* 408 Joint Learned Council for Geological Sciences
       1958-   ?

       Chairman

       Trofimuk, A.A. 1958-

* 409 Joint Learned Council for Geological, Mineralogical, Geophysical and Geographical Sciences
       ?   -1969

       Chairman

       Trofimuk, A.A. . . . . ? -1969

410 Joint Learned Council for Earth Sciences
     1969-

Chairman

    Trofimuk, A.A. 1969-

411 Joint Learned Council for Biological Sciences
       1958-

Chairman

    Dubinin, N.P. 1958-

412 Joint Learned Council for Economic Sciences
       1958-

Chairmen

    Prudenskiy, G.A. 1958-1967
    ? 1967-

INSTITUTIONS OF THE USSR ACADEMY OF SCIENCES' SIBERIAN DIVISION

413 Novosibirsk Institute of Automation and Electrometry
       1957-

Directors

    Karandeyev, K.B. 1957-1968
    Nesterikhin, Yu.Ye. 1968-

414 Institute of Bioactive Compounds, Novosibirsk
       1964-

Directors

    ? 1964-

415 Institute of Biology, Novosibirsk
       1961-

Director

    Cherepanov, A.I. 1961-

* 416 Buryat Comprehensive Research Institute, Ulan-Ude
        1958-

Directors

    Lubsanov, D.D. 1958-1961
    Makeyev, O.V. 1961-

* 417 East Siberian Institute of Biology, near Irkutsk
        1961-1966

Director

    Reymers, F.E. 1961-1966

418 Siberian Institute of Plant Physiology and Biochemistry
        1966-

Director

    Reymers, F.E. 1966-

* 419 East Siberian Institute of Geology, near Irkutsk
        1958-1962

Director

    Odintsov, M.M. 1958-1962

420 Institute of the Earth's Crust
        1962-

Director

    Odintsov, M.M. 1962-

* 421 Institute of Vulcanology, Petropavlovsk (Kamchatka)
1962-

Directors

Piyp, B.I. 1962-1966
Gorshkov, G.S. 1967-1970
Ter-Avanesyan, D.V. 1970-

422 Institute of Siberian and Far Eastern Geography, near Irkutsk
1958-

Directors

Gerasimov, I.P. 1958-1959
Sochava, V.B. 1959-

423 Institute of Geology and Geophysics, Novosibirsk
1957-

Director

Trofimuk, A.A. 1957-

424 Institute of Geochemistry, near Irkutsk
1958-

Directors

Vinogradov, A.P. 1958-1961
Tauson, L.V. 1961-

425 Institute of Hydrodynamics, Novosibirsk
1957-

Director

Lavrent'yev, M.A. 1957-

426 Mining Institute, Novosibirsk
              1957-

              Director

       Chinakal, N.A. 1957-

* 427 Transbaykal Comprehensive Research Institute, Chita
              1961-   ?

              Director

       Mel'nikov, G.A. 1961-   ?

428 Irkutsk Institute of Organic Chemistry
              1958-

              Director

       Shostakovskiy, M.F. 1958-

429 Institute of History, Philology and Philosophy, Novosibirsk
              1965-

              Director

       Okladnikov, A.P. 1965-

430 Institute of Catalysis, Novosibirsk
              1958-

              Director

       Boreskov, G.K. 1958-

* 431 V.N. Sukachyov Timber and Wood Institute, Krasnoyarsk
              1959-

Director

Zhukov, A.B. 1959-

* 432　Limnological Institute, Listvennichnoye, Irkutsk Oblast
　　　1961-

Director

Galaziy, G.I. 1961-

* 433　Mathematics Institute and Computing Center, Novosibirsk
　　　1957-1963

Director

Sobolev, S.L. 1957-1963

434　Mathematics Institute, Novosibirsk
　　　1963-

Director

Sobolev, S.L. 1963-

435　Computing Center, Novosibirsk
　　　1963-

Director

Marchuk, G.I. 1963-

* 436　Institute of Permafrost Studies, Yakutsk
　　　1960-

Director

Mel'nikov, P.I. 1960-

437 Institute of Inorganic Chemistry, Novosibirsk
 1957-

   Director

   Nikolayev, A.V. 1957-

438 Institute of Organic Chemistry, Novosibirsk
 1958-

   Director

   Vorozhtsov, N.N. 1958-

* 439 Institute of Soil Science and Agrochemistry, Novosibirsk
 1967-

   Director

   Kovalyov, R.V. 1967-

* 440 Institute of Radio Physics and Electronics, Sverdlovsk
 1957-1964

   Director

   Rumer, Yu.B. 1957-1964

* 441 Institute of Solid-State Physics and Semiconductor Electronics
 1962-1964

   Director

   Rzhanov, A.V. 1962-1964

442 Institute of Semiconductor Physics
 1964-

Director

  Rzhanov, A.V. 1964-

* 443 Sakhalin Comprehensive Research Institute
    1957-

Directors

  Khel'kvist, G.A. 1957-1963
  Tuyezov, I.K. 1964-

* 444 Northeastern Comprehensive Research Institute
    1960-

Director

  Shilo, N.A. 1960-

* 445 Siberian Institute of Earth Magnetism, Ionosphere and Propagation of Radio Waves, Irkutsk
    1960-

Directors

  Yerofeyev, N.M. 1960-1966
  Stepanov, V.Ye. 1966-

446 Siberian Power Engineering Institute, Novosibirsk
    1960-

Director

  Melent'yev, L.A. 1960-

447 Institute of Theoretical and Applied Mechanics, Novosibirsk
    1957-

Directors

  Khristianovich, S.A. 1957-1965

Zhukov, M.F. 1966
Struminskiy, V.V. 1966-

448 Institute of Thermophysics, Novosibirsk
        1957-

    Directors

    Novikov, I.I. 1957-1965
    Kutateladze, S.S. 1965-

449 Institute of Transport Power Engineering, Novosibirsk
        1961- ?

    Director

    Shcherbakov, V.K. . . . . 1961- ?

* 450 Institute of Physics, Krasnoyarsk
        1957-

    Director

    Kirenskiy, L.V. 1957-1969

* 451 Institute of Metallurgical Chemistry, Novosibirsk
        1962-1964

    Director

    Logvinenko, A.T. 1962-1964

452 Institute of Physical and Chemical Principles of Processing Mineral Raw Materials, Novosibirsk
        1964-

    Director

    Logvinenko, A.T. 1964-

453 Institute of Experimental Biology and Medicine
1957-1965

Directors

Meshalkin, Ye.N. 1957-1963
Borodin, Yu.I. 1963-1965

* 454 Institute of Physiology, Novosibirsk
1965-

Director

Slonim, A.D. 1965-

* 455 Khabarovsk Comprehensive Research Institute
1968-

Director

? 1968-

456 Institute of Chemical Kinetics of Combustion, Novosibirsk
1957-

Director

Koval'skiy, A.A. 1957-

457 Institute of Cytology and Genetics, Novosibirsk
1957-

Directors

Dubinin, N.P. 1957-1960
Belyayev, D.K. 1960-

* 458 Laboratory for Applying Statistical and Mathematical Methods in Economics, Novosibirsk
1958-1967

Directors

Kantorovich, L.V. 1958-1966
Aganbegyan, A.G. 1966-1967

* 459 Economics and Statistics Institute
1957-1958

Director

Nemchinov, V.S. 1957-1958

460 Institute of the Economics and Organization of Industrial Production, Novosibirsk
1958-

Directors

Prudenskiy, G.A. 1958-1966
? 1966-

461 Institute of Nuclear Physics, Novosibirsk
1957-

Director

Budker, G.I. 1957-

462 Central Siberian Botanical Gardens
1962-

Director

Sobolevskaya, K.A. 1962-

* 463 State Public Scientific and Technical Library of the USSR Academy of Sciences' Siberian Division
? -

Director

Kartashev, N.S. 1966-

464　Editorial and Publishing Council, Novosibirsk
1958-

Director

Sobolev, S.L. . . . . 1958-　?

465　Siberian Council of Expeditionary Research, Novosibirsk
1958-

Director

Nekrasov, N.N. . . . . 1958-　?

* 466　Committee for Nature Conservation of USSR Academy of Sciences' Siberian Division
1959-

Chairman

Trofimuk, A.A. 1959-

467　USSR Geographical Society in Siberia and the Far East at USSR Academy of Sciences' Siberian Division, Irkutsk
1959-

Chairmen

?　. . . . . ?　-　?

BRANCHES OF THE USSR ACADEMY OF SCIENCES' SIBERIAN DIVISION

* 468　Buryat Branch
1966-

Chairmen

Makeyev, O.V. 1967-1969
Filippov, V.R. 1969-

* 469   East Siberian Branch
                1957-

        Chairmen

        Krotov, V.A. 1957-1960
        Melent'yev, L.A. 1960-1965
        Odintsov, M.M. 1965-1969
        Sochava, V.B. 1969-

* 470   Far Eastern Branch
                1957-

        Chairmen

        Bykov, V.T. 1957-1960
        Khomentovskiy, A.S. 1960-1966
        Neunylov, B.A. 1966-

* 471   West Siberian Branch
                1957-1959

        Chairman

        Gorbachyov, T.F. 1957-1959

* 472   Yakutian Branch
                1957-

        Chairmen

        Rozhkov, I.S. 1957-1965
        Cherskiy, N.V. 1965-

Notes

1, 2, 3   In 1917 - 1927 the 1836 Statutes of the Russian Academy of Sciences were in force, whereby the Academy was divided into three departments:
          Department of Physical and Mathematical Sciences (1)

4, 5   On 18 June 1927 the USSR Council of Peoples' Commissars approved new Statutes to replace the 1836 Statutes. The Academy was now divided into two departments:
   Department of Mathematical and Natural Sciences (4) and
   Department of Humanitarian Sciences (5).

6, 7, 8   On 23 November 1935 new Statutes of the USSR Academy of Sciences were approved, dividing the Academy into three departments:
   Department of Mathematical and Natural Sciences (6),
   Department of Technical Sciences (7) and
   Department of Social Sciences (8).
The departments consisted of speciality groups.

9 - 20   On 29 September 1938 the USSR Academy of Sciences' General Meeting abolished the existing groups in the Academy's structure and instead of the three departments - Department of Mathematical and Natural Sciences (6), Department of Technical Sciences (7) and Department of Social Sciences (8) - established eight departments:
   Department of Physical and Mathematical Sciences (9),
   Department of Chemical Sciences (10),
   Department of Geological Sciences (11),
   Department of Biological Sciences (12),
   Department of Technical Sciences (13),
   Department of History and Philosophy (14),
   Department of Economics and Law (16),
   Department of Literature and Language (20).
In 1953 the Departments of History and Philosophy (14) and of Economics and Law (16) were reorganized into the Department of Historical Sciences (15) and the Department of Economics, Philosophy and Law (17).

In 1962 the Department of Economics, Philosophy and Law (17) was divided into two departments: the Department of Philosophical and Law Sciences (18) and the Department of Economic Sciences (19).

26	On 7 March 1968 the Department of Earth Sciences (26) was reorganized into the Earth Sciences Section, consisting of two departments (28 and 29).
39	The All-Union Institute of Scientific and Technical Information was organized in 1952 as the "Institute of Scientific Information" and placed under the USSR Academy of Sciences' Publishing House. In 1955 renamed the All-Union Institute of Scientific and Technical Information.
40	The Library of the Academy of Sciences was founded in 1714 as an independent institution. First records of books issued date from 1718. In 1725 the Library was placed under the Academy of Sciences. In 1742 the first printed catalogue appeared. The USSR Academy of Sciences' Library is the main library center of all the Academy's libraries.
42	The Editorial and Publishing Council, which is always chaired by the Academy's President, existed already in 1936. The exact date of its founding is unknown.
43	The Commission for the Study of Natural Production Resources of Russia was founded on 4 February 1915 by a group of academicians headed by V. I. Vernadskiy. Its action program included the publication of a joint work describing known natural resources and special research on little known natural objects. In 1930 the commission was reorganized and renamed the Council for the Study of Production Resources of the USSR.
44	The Council for the Study of Production Resources of the USSR was in 1960 transferred to the administration of the USSR Council of Ministers' Scientific Economic Council.
52	The Physics Laboratory was founded in 1912. In 1921 it was amalgamated with the Mathematics Center and renamed the Institute of Physics and Mathematics (53).

53	The Institute of Physics and Mathematics was founded in 1921 by Academician V.A. Steklov. In 1934 it was divided into the V.A. Steklov Mathematics Institute (54) and P.N. Lebedev Physics Institute (55).
56	The Acoustics Commission's founding date is unknown. In 1938 it was transferred to the Institute of Physics (55). In 1965 the commission was reorganized into the Scientific Council for Acoustics (120).
58	Established in 1959 from the Computer Center of the Leningrad Department of the USSR Academy of Sciences' Mathematical Institute.
60	The Sverdlovsk Institute of Mathematics and Mechanics was established in 1970 from the V. A. Steklov Institute of Mathematics' Sverdlovsk Department.
63	According to the CC, CPSU and USSR Council of Ministers' decree on "Measures to Improve the Coordination of Research Work in Our Country and the USSR Academy of Sciences' Activity," the Acoustics Institute was in April 1961 transferred to branch departmental administration.
65	In 1958 the laboratory was reorganized as the Institute of High-Pressure Physics (66).
67	The Institute of Metal Physics was in 1958 withdrawn from the jurisdiction of the USSR Academy of Sciences' Ural Branch and became an independent institute directly subordinate to the USSR Academy of Sciences' Department of Physical and Mathematical Sciences.
71	In 1956 the institute was incorporated into the Dubno Joint Nuclear Research Institute.
72	The Institute of Atomic Energy was in 1963 transferred to branch departmental administration.
73	The Radium Institute was founded in 1922 by

Academician V. I. Vernadskiy. Until 1938 it was a state institute, not incorporated into the USSR Academy of Sciences. In 1961 the institute was transferred to branch departmental jurisdiction according to the CC, CPSU and USSR Council of Ministers' Decree on "Measures to Improve the Coordination of Research Work in Our Country and the USSR Academy of Sciences' Activity."

75   The Radio Engineering Institute was in 1961 transferred to branch departmental jurisdiction according to the CC, CPSU and USSR Council of Ministers' Decree on "Measures to Improve..."

76   The Institute of Physical Engineering was founded in 1918 from the State Institute of Roentgenology and Radiology's Department of Physical Engineering. In 1938 the institute was incorporated into the USSR Academy of Sciences. The institute is located in Leningrad.

77   In 1963 the Kazan' Branch of the USSR Academy of Sciences was abolished and the institute incorporated into the Academy.

78   Established in 1930 from the Commission for the Study of Natural Production Resources of the USSR's (44)Power Engineering Department. In 1961 transferred to branch departmental jurisdiction according to the CC, CPSU and USSR Council of Ministers' Decree on "Measures to Improve..."

79   The Semiconductors Laboratory was in 1954 reorganized into the Semiconductors Institute (80).

81   The High Temperatures Laboratory, founded in 1960, was in 1961 transferred to branch departmental jurisdiction according to the CC, CPSU and USSR Council of Ministers' Decree on "Measures to Improve..."

82   The High Temperatures Institute was in 1967 reincorporated into the Academy as the Re-

search Institute of High Temperatures (81). The High Temperatures Institute was founded in 1967 from the departmental Research Institute of High Temperatures, formerly established from the High Temperatures Laboratory (81).

84 The Institute of Comprehensive Transport Problems was in 1960 transferred to the administration of the USSR Council of Ministers' Scientific Council for the Economy.

86 The Institute of Machine Science was in 1961 transferred to branch departmental administration according to the CC, CPSU and USSR Council of Ministers' Decree on "Measures to Improve..."

87 The Engine Laboratory was in 1961 transferred to branch departmental administration according to the CC, CPSU and USSR Council of Ministers' Decree on "Measures to Improve..."

88 The Laboratory of Electric Welding Machines was in 1955 detached from the Institute of Automation and Telemechanics' (93) Leningrad Department as an independent laboratory. In 1961 it was transferred to branch departmental administration according to the CC, CPSU and USSR Council of Ministers' Decree on "Measures to Improve..."

89 The Laboratory of Control Machinery and Systems was in 1958 reorganized into the Institute of Electronic Control Machinery (90). The date of the laboratory's establishment is unknown.

90 The Institute of Electronic Control Machinery was founded on the basis of the Laboratory of Control Machinery and Systems (89) in 1958. In 1961 it was transferred to branch departmental administration according to the CC, CPSU and USSR Council of Ministers' Decree on "Measures to Improve..."

91 The Commission for Telemechanics and Automa-

tion was founded in 1934. In 1936 it was reorganized into a Permanent Commission and incorporated into the Department of Technical Sciences. In 1938 the USSR Academy of Sciences' Presidium saw fit to replace this commission with the Automation Committee, founded under the Academy's Presidium.

94 The Institute of Electromechanics was founded in 1956 from the Institute of Automation and Telemechanics' Leningrad Department. In 1961 it was transferred to branch departmental administration according to the CC, CPSU and USSR Council of Ministers' Decree on "Measures to Improve..."

95 The Laboratory for the Scientific Development of Wire Communications Problems was in 1948 detached from the Institute of Automation and Telemechanics. In 1959 it was renamed the Laboratory of Data Transmission Systems (96).

96 The Laboratory of Data Transmission Systems was in 1962 reorganized into the Institute of Data Transmission Problems (97).

98 The Institute of Fine Mechanics and Computer Engineering was founded from the Institute of Machine Science's Department of Fine Mechanics, the Mathematical Institute's Department of Approximate Computation and the Power Institute's Laboratory for Electromodelling and Electrometry. In 1961 the institute was transferred to branch departmental administration according to the CC, CPSU and USSR Council of Ministers' Decree on "Measures to Improve..."

99 The Institute of Mechanics was in 1965 reorganized into the Institute of Mechanical Problems (100).

101 The Main Astronomical Observatory, the former Pulkovo Observatory, was in 1934 transferred from the RSFSR People's Commissariat of Education to the USSR Academy of Sciences. The Observatory was established in 1839.

102	The Commission for Solar Research was in 1934 incorporated into the Main Astronomical Observatory (101).
104	The Crimean (Simeiz) Astrophysical Observatory was established in 1908 as the Pulkovo Observatory's Simeiz Department. In 1945 it served as the basis for the Crimean Astrophysical Observatory.
106	The Astronomical Institute was founded in 1923 from the State Computation Institute. In 1939 it was incorporated into the USSR Academy of Sciences. In 1943 it was reorganized into the Institute of Theoretical Astronomy (107).
108	The Institute of Earth Magnetism, Ionosphere and the Propagation of Radio Waves was in 1959 removed from the Ministry of Communications' jurisdiction and incorporated into the USSR Academy of Sciences.
110	The Spectroscopic Commission was founded in 1940. The Institute of Spectroscopy was founded from this commission in 1969 (111).
127	This institution developed from Peter I's Chamber of Arts in 1724 and was variously known in the history of Russian geology as the Mineral Cabinet, Mineralogical Museum, Geological Museum and Museum of Mineralogy and Geology. In 1925 the museum was split into the Geological and the Mineralogical Museums (132). The museum's Geological Department was simultaneously renamed the "Peter I Geological Museum" (128).
128	In 1930 the Peter I Geological Museum was divided into three independent institutes: Geological Institute (129), Paleozoological Institute (130) and the Petrographical Institute (131).
129	Under the USSR Council of People's Commissars' decree of 17 November 1937 the Geological Institute was amalgamated with the Petrographical Institute (131) and the Insti-

	tute of Geochemistry, Mineralogy and Crystallography (133) and renamed the Institute of Geological Sciences (134).
132	The Museum of Mineralogy was in 1932 incorporated into the M.V. Lomonosov Institute of Geochemistry, Mineralogy and Crystallography (133).
133	Founded in 1932 from the Museum of Mineralogy, the Crystallography Laboratory (136) and the Geochemical Department of the former Commission for the Study of Natural Production Resources of the USSR. In 1937 amalgamated with the Geological Institute (129) and Petrographical Institute (131) to form the Institute of Geological Sciences (134).
135	The Geological Institute was in 1955 reorganized from the Institute of Geological Sciences (134).
136	The Crystallography Laboratory, until 1932 an independent institution, was then incorporated into the Institute of Geochemistry, Mineralogy and Crystallography (133). In 1938 it again became independent.
137	In 1955 detached from the Geological Institute (135).
138	Founding date unknown. In 1967 the laboratory was expanded into the Institute of Pre-Cambrian Geology and Geochronology (139).
140	Founding date unknown. In 1956 it served as the basis for the Institute of Mineralogy, Geochemistry and Crystallochemistry of Rare Elements (141), which in 1961 was transferred to branch departmental administration.
142	The Geochemical Problems Laboratory was founded in early 1929 by Academician V. I. Vernadskiy. In 1947 it was expanded into the V. I. Vernadskiy Institute of Geochemistry and Analytical Chemistry (143).

144   The Sapropelite Institute was founded in 1932. In December 1934 it was reorganized into the Institute of Mineral Fuels (145).

145   The Institute of Mineral Fuels was founded in 1934 by I. M. Gubkin, who was also its first director. It was founded from the Sapropelite Institute and some laboratories of the Oil Research Institute and the Oil Prospecting Institute.

146   In 1947 detached from the Institute of Mineral Fuels.

147   The Institute of Geology and Exploitation of Mineral Fuels was founded in 1959 from the Institute of Mineral Fuels and the Oil Institute. In 1961 it was transferred to branch departmental administration according to the CC, CPSU and USSR Council of Ministers' Decree on "Measures to Improve..."

148   The Mining Institute was founded in 1938 from the Mining Group of the USSR Academy of Sciences' Department of Technical Sciences, which had existed from 1935. In 1961 it was transferred to branch departmental administration according to the CC, CPSU and USSR Council of Ministers' Decree on "Measures to Improve..."

150   In 1961 transferred to branch departmental administration.

151   The Meteorites Committee was founded in 1939 as a result of reorganization of the USSR Academy of Sciences' Meteorites Commission.

153   The Geomorphological Institute was founded in 1930 from the former Geographical Department of the Commission for the Study of Natural Production Resources of the USSR. In 1933 it was incorporated into the Institute of Sandy Deserts (154). In 1934 it served as the basis for the Institute of Physical Geography (155).

155   The Institute of Physical Geography was in

1936 reorganized into the Institute of Geography (156).

157 The Commission (from 1936 Committee) for Permafrost Studies was in 1939 reorganized into the Institute of Permafrost Studies (158).

158 In 1960 the Northeastern Department of the Institute of Permafrost Studies, the Igar, Aldan and Anadyr Permafrost Stations were incorporated into the Institute of Permafrost Studies of the USSR Academy of Sciences' Siberian Division.

159 The Institute of Theoretical Geophysics was in 1947 amalgamated with the Seismological Institute (161) to form the Geophysical Institute (162).

160 The Permanent Central Seismic Commission has existed since 1900. In 1897 the Seismological Committee of the British Association offered to collaborate with the Russian Academy of Sciences in recording earth tremors from remote earthquakes. The Russian Academy of Sciences accepted the offer and in 1897 founded a Provisional Seismic Commission which in 1900 became a Permanent Commission. Together with the Physics Institute's now separate Seismological Commission, it served in 1928 as the basis for the Seismological Institute (161).

161 The Seismological Institute was in 1947 amalgamated with the Institute of Theoretical Geophysics (159) into the Geophysical Institute (162).

162 The Geophysical Institute was in 1956 divided into three institutes: the Institute of Geophysics (163), the Institute of Atmospheric Physics (164) and the Institute of Applied Geophysics (165).

165 The Institute of Applied Geophysics was in 1961 transferred to branch departmental administration according to the CC, CPSU and

	USSR Council of Ministers' Decree on "Measures to Improve..."
166	Founding date unknown. In 1962 incorporated into the Institute of Vulcanology of the USSR Academy of Sciences' Siberian Division.
167	Founding date unknown. In 1948, together with the Black Sea Hydrophysical Station, it was reorganized into the Marine Hydrophysical Institute (168), which in 1961 was transferred to branch departmental administration.
169	The Oceanological Laboratory was founded in 1941 to promote comprehensive oceanological research. In 1946 it was reorganized into the Institute of Oceanology (170).
171	Founding date unknown. The earliest records date from 1946. In 1969 the laboratory was transferred from Leningrad University to the USSR Academy of Sciences.
173	Founding date unknown. In 1953 it was incorporated into the USSR Academy of Sciences' East-Siberian Branch. In 1961 it formed the basis for the Limnological Institute of the USSR Academy of Sciences' Siberian Division.
174	The V. V. Dokuchayev Soil Institute was founded in 1925 from the Soil Committee and the Commission for the Study of Russian Production Resources' Museum. In 1961 it was transferred to branch departmental administration according to the CC, CPSU and USSR Council of Ministers' Decree on "Measures to Improve..."
175	The V. V. Dokuchayev Central Museum of Soil Science was founded in 1947 as an independent institution to commemorate V.V. Dokuchayev. The museum was originally established in 1904 by the Free Economic Society, in 1918 incorporated into the Academy of Sciences, and in 1925 transferred to the Soil Institute. After the institute's transfer to Moscow in 1934 the museum continued in Leningrad.

176 The Tectonics Commission was founded in 1948. In 1961 it served as the basis for the Commission of International Tectonic Charts (177).

178 The Committee of Geodesy and Geophysics was founded in 1955 in connection with the USSR's joining UNESCO's International Geodesical and Geophysical Union.

179 The Antarctic Research Council was founded in 1955 and replaced the former Joint Committee for the International Geodesic Year's Commission for Antarctic Research. In 1959 the council was abolished and the Interdepartmental Commission for Antarctic Research established (180).

182 The Scientific Council for Soil Science and Melioration Problems was in 1968 reorganized from the Scientific Council for the "Theoretical Foundations of Soil Science."

186 The Institute of Organic Chemistry was founded in 1934 from the Laboratories of Organic Synthesis and High Pressures which existed from 1929-30.

187 Founding date unknown. In 1934 it was incorporated into the Institute of Organic Chemistry (186).

188 The A. Ye. Arbuzov Chemical Institute in Kazan' was in 1962 transferred from the Kazan' Branch to the direct administration of the USSR Academy of Sciences. In 1965 it was amalgamated with the Kazan' Institute of Organic Chemistry (189) to form the A. Ye. Arbuzov Institute of Organic and Physical Chemistry in Kazan' (190).

191 In 1934 reorganized into the Colloidal and Electrochemical Institute (192).

192 In 1945 served as the basis for the Institute of Physical Chemistry (193).

194 Founded in 1931 from the Physical Chemistry

	Department of the Leningrad Physical-Engineering Institute.
195	Founding date unknown. In 1959 incorporated into the Institute of Chemical Physics (194).
198	Founded in 1939. In 1961-64 under branch departmental administration according to the CC, CPSU and USSR Council of Ministers' Decree on "Measures to Improve..."
199	The Magnetics Laboratory was detached from the Metallurgical Institute (198) in 1957. In 1961 it was transferred to branch departmental administration.
201	Founded by M. V. Lomonosov in 1748 and named the Central Chemical Laboratory, it was Russia's first chemical laboratory. After a fire in 1867 it was reestablished and named the Laboratory of General Chemistry. In 1934 it was amalgamated with the Institute of Physical and Chemical Analysis (202) and the Platinum Institute into a joint Institute of General and Inorganic Chemistry (204).
202	The Institute of Physicochemical Analysis was founded in 1917. Originally run by the Commission for the Study of Natural Production Resources. In 1934 it was incorporated into the Institute of General and Inorganic Chemistry (204).
205	Founded in 1948 from the Laboratory of Silicate Chemistry, which existed from 1934. In 1962 it was named for Academician I. V. Grebenshchikov. Located in Leningrad.
206	Founded in 1963 in Gorky. In 1969 it was reorganized as the Institute of Chemistry, Gorky (207).
208	The Novocherkassk Hydrochemical Institute was founded in 1921. In 1938 it was incorporated into the USSR Academy of Sciences. In 1961 it was transferred to branch departmental administration.

210 The Institute of Biological Physics in Pushchino, Moscow Oblast, was founded in 1952 from the Laboratory of Isotopes and Radiation Biophysics, which existed from 1950.

211 Founding date unknown. In 1934 it was reorganized into the Biochemical Institute (213) and the Laboratory of Animal Physiology (212).

212 The Laboratory of Animal Physiology was in 1936 incorporated into the I.P. Pavlov Physiological Institute.

213 In 1944 the institute was named for A. N. Bakh.

215 The Microbiological Laboratory was in 1934 reorganized into the Institute of Microbiology (216).

217 The Institute of Radiation and Physicochemical Biology was in 1965 renamed the Institute of Molecular Biology (218).

219 Founded as an Electronic Microscopy Center but soon renamed a laboratory. In 1967 it was incorporated into the Institute of Molecular Biology (218).

220 Developed from a small Botanical Laboratory founded by Academician A. S. Famintsyn in 1890. In 1922 it was renamed the Laboratory of Plant Biochemistry and Physiology. In 1934 it was transferred to Moscow and reorganized into the K. A. Timiryazev Institute of Plant Physiology (221).

221 In 1936 named for K.A. Timiryazev.

223 The Protein Institute located at the USSR Academy of Sciences' Scientific Center for Biological Research in Pushchino, Moscow Oblast, was founded in 1967.

225 Founded in 1889, when a small physiological laboratory established by Academician F. V. Ovsyannikov in 1864 became an independent

	institution. After Ovsyannikov's death in 1907 I. P. Pavlov succeeded as director. In 1925 the laboratory was renamed the Physiological Institute (226).
226	The Physiological Institute was in 1934 renamed the Institute of Physiology and Pathology of Higher Nervous Activity (227). After I. P. Pavlov's death in 1936 the institute was amalgamated with the Laboratory for Animal Physiology (212) and was renamed the I.P. Pavlov Physiological Institute.
228	The I. P. Pavlov Institute of Physiology was founded in 1950 from the I.P. Pavlov Physiological Institute (226) and the Institutes of Physiology of the Central Nervous System and USSR Academy of Medical Sciences' I. P. Pavlov Institute of Evolutionary Physiology and Pathology of Higher Nervous Activity.
229	The Institute of Higher Nervous Activity was founded in 1950 according to a resolution of a joint session of the USSR Academy of Sciences and Academy of Medical Sciences on problems of I.P. Pavlov's physiological theories. In 1960 this institute was amalgamated with the Physiological Laboratory (225) and renamed the Institute of Higher Nervous Activity and Neurophysiology (231).
232	Founding date unknown. On 10 June 1948 it was transferred to USSR Academy of Medical Sciences' administration.
233	The Laboratory for the Study of Nervous and Humoral Regulation was founded in 1963 by amalgamating the USSR Academy of Medical Sciences' Laboratory of Clinical Neurophysiology with the USSR Academy of Sciences' Laboratory of Neurohumoral Regulation. In 1969 it was reorganized into the N. I. Grashchenkov Laboratory of Control Problems in Human and Animal Organisms (234).
235	In 1956 the laboratory was reorganized into the I. M. Sechenov Institute of Evolutionary Physiology (236).

236	In 1964 renamed the Institute of Evolutionary Physiology and Biochemistry (237).
238	The N. M. Morozov "Borok" Biological Station was in 1947 reorganized from the "Borok" Biological Ward into an independent station. In 1956 it was reorganized into the Institute of Water Reservoirs Biology (239).
239	In 1962 renamed the Institute of Inland Waters Biology (240).
241	In December 1934 amalgamated with Paleozoological Institute (242) into the Institute of Evolutionary Morphology and Paleozoology (243).
243	The Institute of Evolutionary Morphology and Paleozoology was in 1937 divided into two institutes: Institute of Paleontology (244) and A.N. Severtsov Institute of Evolutionary Morphology (245).
245	The A.N. Severtsov Institute of Evolutionary Morphology was in 1949 amalgamated with the Institute of Cytology, Hystology and Embryology (247) into the A.N. Severtsov Institute of Animal Morphology (248).
246	The Institute of Experimental Biology was in 1938 transferred to the USSR Academy of Sciences. In 1940 it was reorganized into the Institute of Cytology, Hystology and Embryology (247).
247	In 1949 incorporated into the A.N. Severtsov Institute of Animal Morphology (248).
248	The A. N. Severtsov Institute of Animal Morphology in 1967 was divided into two institutes: A. N. Severtsov Institute of Evolutionary Morphology and Ecology of Animals (249) and Institute of Biological Development (250).
252	The Botanical Museum was founded in 1824 by the Petersburg Academy of Sciences. In 1931 it was amalgamated with the Main Botanical

Gardens (253) into the Botanical Institute, which in 1940 was named for V. L. Komarov (254).

253 The Main Botanical Gardens were founded by a decree of Peter I in 1714 on Apothecary's Island near Petersburg. Originally named the "Apothecary's Kitchen-Garden," they were later renamed the "Medical Gardens." In 1823 they were renamed the Botanical Gardens, after the Revolution the RSFSR Main Botanical Gardens and then the USSR Main Botanical Gardens. In 1929 they were incorporated into the USSR Academy of Sciences, and in 1931 amalgamated with the Botanical Museum into the USSR Academy of Sciences' Botanical Institute.

256 The Forest Institute was founded by Academician V.N. Sukhachyov. In 1959 it was transferred to Krasnoyarsk and incorporated into the USSR Academy of Sciences' Siberian Division.

257 The Forestry Laboratory was founded in 1959 to continue the comprehensive research in the European USSR begun by the Forest Institute (256). In 1970 it was reorganized into the Institute of Forestry (258).

259 The Institute of Timber and Timber Chemistry, Arkhangel'sk developed from the Northern Branch of the USSR Academy of Sciences' Timber Institute in Arkhangel'sk.

260 The Genetics Laboratory was in 1934 reorganized into the Institute of Genetics (261).

261 The Institute of Genetics was in 1966 renamed the Institute of General Genetics (262).

263 The Zoological Museum was founded in 1832. The museum's collection was based on the zoological specimens of Peter I's Chamber of Arts, founded in 1714. In 1931 the museum was incorporated into the USSR Academy of Sciences' Zoological Institute in Leningrad (264).

265	The Sevastopol' Biological Station was founded in 1871-72. In 1938, since the station was actually a hydrobiological station, it was decided to incorporate it into the Zoological Institute. During WW II the station was destroyed and reconstruction work did not begin until 1945, although in 1944 the station again became an independent institution. In 1957 it was reorganized into the Institute of South Sea Biology (266).
268	Founding date unknown. In 1938 it was incorporated into the Zoological Institute (264), but in 1949 mentioned as an independent institution. In 1953 the station was incorporated into the Academy's Kola Branch.
269	The Scientific Directors' Council of the Biological Research Center in Pushchino was founded in 1962 according to a USSR Academy of Sciences' Presidium resolution. In Pushchino there has arisen a complex of biological research institutes. It includes: Institute of Natural Compounds Chemistry (222), Institute of General Virology, Institute of Microbiology (216), Institute of Biophysics, Institute of Plant Physiology (221), Institute of Animal Biochemistry.
286	The Institute of Oriental Studies was founded in 1930 from the Asiatic Museum (282), the Institute of Buddhist Culture (283), the Turcological Study Center (284) and the Orientalists Collegium. In 1960 it was amalgamated with the Sinological Institute (287) into the Institute of Asian Peoples (289). In 1968 it was again renamed the Institute of Oriental Studies. It is the main center for the history, economics, literature and languages of the Orient. In 1950 the Pacific Institute (285) was also incorporated into this institution.
289	The institute studies the economics, history, sociology and culture of African peoples. It concentrates on topical problems, such as the campaign to liquidate colonial regimes, establish national states and enforce their sovereignty.

290   The Scientific Council for African Problems originated from the Commission for the Study of Africa in 1966.

293   In 1836 from the Chamber of Arts there were detached the Ethnographical and the Anatomical Museums, which in 1878 were amalgamated into the Museum for Anthropology and Ethnography. In 1933 this museum and the Institute for the Study of the Peoples of the USSR (295) were incorporated into the Institute of Anthropology, Archeology and Ethnography (296).

294   On the basis of this commission in 1930 there was founded the Institute for the Study of the Peoples of the USSR (295).

296   In 1937 reorganized into the Institute of Ethnography (298).

297   In 1936 incorporated into the Institute of Anthropology, Archeology and Ethnography (296). In 1954 the museum was renamed the Museum of the History of Religion and Atheism.

298   In 1943 the institute was transferred from Leningrad to Moscow, but a Leningrad Branch with its Museum of Anthropology and Ethnography was retained. In 1947 it was named for N.N. Miklukho-Maklay.

299   On 18 April 1919 there was founded the Russian Academy of Material Culture History from the Archeological Commission which had existed from 1859. In 1926 the academy was renamed the State Academy of Material Culture and in 1937 it was incorporated into the USSR Academy of Sciences as the Institute of Material Culture History, sited in Leningrad but with a branch in Moscow. In 1945 the Moscow branch became the institute's center, with a department in Leningrad. In 1959 it was reorganized into the Institute of Archeology (300).

301   The History of Knowledge Commission was

founded in 1926 by Academician V. I. Vernadskiy. In February 1932 it was renamed the Institute for the History of Science and Engineering (302).

302 The Institute of the History of Science and Engineering was in 1936 reinforced by the Commission of Engineering History of the Communist Academy's Presidium. On 29 April 1938 this institute was abolished, but the Commission of Engineering History remained as an independent institution. In 1953 the commission was incorporated into the Institute of the History of Natural Science and Engineering (305).

303 In 1953 the Commission for the History of Engineering was incorporated into the Institute of the History of Natural Science and Engineering (305).

305 The following commissions were incorporated into this institute: Engineering History (303), Chemical Sciences (307), Geological and Geographical Sciences (306), Development and Publication of A.M. Butlerov's Scientific Legacy (308), Development and Publication of M.V. Lomonosov's Scientific Legacy (309), Development and Publication of D.I. Mendeleyev's Scientific Legacy (310), the Lomonosov Museum (311).

313 Transferred to branch departmental administration according to the CC, CPSU and USSR Council of Ministers' Decree on "Measures to Improve..."

314 The Archeographic Commission was founded in 1834 in Petersburg at the Ministry of Public Education. In 1837 it was confirmed as a permanent institution for systematically publishing national history sources. In 1922 it was incorporated into the USSR Academy of Sciences and in 1926 amalgamated with the Permanent Historical Commission into the USSR Academy of Sciences' Archeographic History Commission. In 1944-48 it was placed under Historical Institute (318) and from

1956 functioned as an independent institution (314).

316   The Institute of the History of Archeography (314) was founded in 1931 from the Archeographic History Commission (315). In 1936, in connection with the amalgamation of the Communist Academy with the USSR Academy of Sciences' institutions, it was reorganized into the Institute of History (318).

318   The Institute of History was founded from the Communist Academy's Institute of History and the USSR Academy of Sciences' Institute of the History of Archeography (316). In 1968 it was divided into two institutes: Institute of the History of the USSR (319) and the Institute of Universal History (320).

321   The Institute of Slavic Studies was founded in 1947 from the Academy of Sciences' Slavic Studies Section, which had existed from 1942, and the Slavic Commission. In 1969 it was renamed the Institute of Slavic and Balkan Studies (322).

323   The Institute of Philosophy was founded in 1936 from the Communist Academy's Institute of Philosophy.

324   The Institute of Specific Social Research was founded in 1968 from the Department of Specific Sociological Research of the USSR Academy of Sciences' Institute of Philosophy (323).

325   The Institute of State Law was founded in 1936 from the Communist Academy's Institute of Soviet Construction and Law. In 1938 it was reorganized into the Institute of Law (326).

326   In 1960 reorganized into the Institute of State and Law (327).

328   Transferred in 1936 from the Communist Academy. In 1947 it was amalgamated with the USSR Academy of Sciences' Institute of Economics.

329   The Institute of Economics was founded in 1936 from the Communist Academy's Agrarian and Economic Institutes. In 1947 it was amalgamated with the Institute of World Economics and World Politics (328).

333   The Central Economic Mathematics Institute was founded in 1963 from the Laboratory of Economic Mathematical Methods (332), the Department of Economic Mathematics of the Computing Center's Programming Laboratory, the Mathematical Group of the Institute of Economics' (329) Department of the Economic Efficacy of Capital Investment and the scientific institutions of the USSR Gosplan.

336   The A.M. Gorky Institute of World Literature was founded in 1932 in connection with the 40th anniversary of the start of Gorky's literary career. In 1938 it was incorporated into the USSR Academy of Sciences.

337   The Pushkin House was founded in Leningrad in 1905. In 1907 it was incorporated into the Academy of Sciences. In 1930 it was reorganized into the Institute of New Russian Literature (338).

338   The Institute of New Russian Literature was in 1932 renamed the Institute of Russian Literature (339).

340   The Caucasian Institute of the History of Archeology was founded in 1917. When Caucasian sources ceased to be the sole material for Japhetic research there arose the need for a new institute for the study of other languages. On N. Ya. Marr's initiative such an institute was founded in September 1921 and named the Institute of Japhetic Studies (341). In 1922 it was renamed the Japhetic Institute (342). The Institute of Language and Thinking was founded from it in late 1931 (343).

343   The N. Ya. Marr Institute of Language and Thinking was in 1950 amalgamated with the Institute of Russian Language (344) into the

	Institute of Linguistics (345).
344	In 1958 the Institute of Russian Language was reestablished as an independent institution.
346	The Institute of Speech and Written Language was founded in April 1938 from the Central Research Institute of Speech and Written Languages of the Peoples of the USSR.
347	In 1963, when the USSR Academy of Sciences' Kazan' Branch was abolished, the Kazan' Institute of Language, Literature and History was transferred to the direct administration of the USSR Academy of Sciences. In 1967 the institute was named for Galimdzhan Ibragimov.
348	In 1963, when the USSR Academy of Sciences' Karelian Branch was abolished, the Petrozavodsk Institute of Language, Literature and History was transferred to the direct administration of the USSR Academy of Sciences. In 1967 the Karelian Branch was reestablished and the institute returned to this branch.
349	In 1963, when the USSR Academy of Sciences' Bashkir Branch was closed, the Ufa Institute of History, Language and History passed to USSR Academy of Sciences' administration. In 1967 the Bashkir Branch was reestablished and the institute returned to this branch.
352	The Commission coordinates research on the history of world culture and directs Soviet scientists' work at the UNESCO Commission.
369	The Archives were first mentioned in 1728. Until 1922 they were named the Conference Archives (of the General Assembly). In 1922 the Conference Archives were reorganized as the All-Academy Archives. Further reorganization in 1963 transferred the Archives Center to Moscow, with a department remaining in Leningrad.

370 The Main Library of Social Sciences was founded in Moscow in 1918 at the Socialist Academy of Social Sciences. In 1924 it was renamed the Library of the USSR Central Executive Committee's Communist Academy. In 1936 it was incorporated into the USSR Academy of Sciences.

371 The Far Eastern Branch was founded in 1932 from the Vladivostok Mountain and Taiga Station. The branch was named for V.L. Komarov. In 1957 this branch passed into the administration of the USSR Academy of Sciences' Siberian Division.

372 The Transcaucasian Branch was founded in 1932 in Tiflis. In 1934 it was divided into three branches: Azerbaydzhani Branch (373), Armenian Branch (374) and Georgian Branch (375).

373 In 1945 the branch was reorganized into the Azerbaydzhani Academy of Sciences.

374 In 1943 the branch was reorganized into the Armenian Academy of Sciences.

375 In 1941 the branch was reorganized into the Georgian Academy of Sciences.

376 The USSR Academy of Sciences' Ural Branch in Sverdlovsk is a joint research institution for studying the Urals' natural resources and developing methods for their exploitation.

377 In 1935 reorganized into the USSR Academy of Sciences' Kazakh Branch in Alma Ata (378).

378 In 1946 the branch was reorganized into the Kazakh Academy of Sciences.

379 The USSR Academy of Sciences' Tadzhikistani Research Base was founded in 1932 in Stalinabad (now Dushanbe). In 1941 it was reorganized into the USSR Academy of Sciences' Tadzhik Branch (380).

380	In 1951 reorganized into the Tadzhik Academy of Sciences.
381	The USSR Academy of Sciences' S.M. Kirov Kola Research Base was founded in 1934 from the USSR Academy of Sciences' Mountain Research Station, which existed from 1930 in Kirovsk. In 1950 it was reorganized into the USSR Academy of Sciences' Kola Branch (382).
384	The Komi Research Base was founded in 1944 in Syktyvkar. In 1949 it was reorganized into the USSR Academy of Sciences' Komi Branch.
386	The USSR Academy of Sciences' Uzbek Branch was founded in 1940 from the Uzbek Committee for Management of Research Institutions which existed from 1932. In 1943 the branch was reorganized into the Uzbek Academy of Sciences.
387	In 1951 reorganized into the Turkmen Academy of Sciences.
388	In 1954 reorganized into the Kirghiz Academy of Sciences.
389	In 1957 incorporated into the USSR Academy of Sciences' Siberian Division.
390	The USSR Academy of Sciences' Kazan' Branch was founded in 1945 from the Kazan' Research Institute of Language, Literature and History. In 1963 the branch was abolished. A number of institutions were transferred to other administrations; into the USSR Academy of Sciences were incorporated the Physical Engineering Institute and the Institute of Language, Literature and History (now the Kazan' Institute of Physical Engineering (77) and the Kazan' Institute of Language, Literature and History (347).
391	Abolished in 1963. A number of institutions were transferred to other administrations. Into the USSR Academy of Sciences were incorporated: Institute of History, Language

and Literature (349) and the Bashkir Branch's Department of Economic Research. In 1967 the Bashkir Branch was reestablished.

392 The USSR Academy of Sciences' East Siberian Branch was founded in 1949 in Irkutsk. In 1957 it was incorporated into the USSR Academy of Sciences' Siberian Division.

393 In 1950 reorganized into the USSR Academy of Sciences' Daghestani Branch (394).

395 In 1949 reorganized into the USSR Academy of Sciences' Karelo-Finnish Branch (396).

396 In 1956 the Karelo-Finnish Branch was renamed the USSR Academy of Sciences' Karelian Branch (397).

397 The USSR Academy of Sciences' Karelian Branch ceased to exist in 1963. A number of its institutions were transferred to other administrations. Into the USSR Academy of Sciences were incorporated: Petrozavodsk Institute of Language, Literature and History (348) and the White Sea Biological Station. In 1967 the Branch was reestablished.

398 In 1949 reorganized into the USSR Academy of Sciences' Yakutian Branch (399), which in 1957 was incorporated into the USSR Academy of Sciences' Siberian Division.

400 In 1949 reorganized into the USSR Academy of Sciences' Sakhalin Branch (401), which in 1955 was renamed the USSR Academy of Sciences' Sakhalin Comprehensive Research Institute.

402 In 1949 reorganized into the USSR Academy of Sciences' Moldavian Branch (403).

403 In 1960 reorganized into the Moldavian Academy of Sciences.

404 Located in Simferopol'. In 1949 reorganized into the USSR Academy of Sciences' Crimean Branch (405).

405   Abolished in 1954, after the incorporation of the Crimea into the Ukrainian SSR.

408   Later renamed the Joint Learned Council for Geological, Mineralogical, Geophysical and Geographical Sciences (409).

409   In 1969 renamed the Joint Learned Council for Earth Sciences (410).

416   Founded in 1958 from the Buryat-Mongolian Institute of Culture and the Buryat-Mongolian Group of the USSR Academy of Sciences' East Siberian Branch.

417   In 1967 reorganized into the Siberian Institute of Plant Physiology and Biochemistry (418).

419   In 1962 renamed the Institute of the Earth's Crust (420).

421   Founded in 1962 from the Kamchatka Geological and Physical Observatory and the USSR Academy of Sciences' Vulcanological Laboratory, which had been transferred to Kamchatka.

427   Founded in 1961 from the Chita Comprehensive Laboratory.

431   Founded in 1959 from the USSR Academy of Sciences' Forest Institute (256), which was transferred to Krasnoyarsk. In 1967 named for V.N. Sukachyov.

432   Founded from the Baykal Limnological Station of the USSR Academy of Sciences at Listvennichnoye, Irkutsk Oblast.

433   In 1963 the Computing Center was detached and made an independent institution. The institute was then divided into two separate institutions: Mathematics Institute (434) and Computing Center (435).

436   Founded in 1960 from the Northeastern De-

partment of the USSR Academy of Sciences' Institute of Permafrost Studies and the Igara, Aldan and Anadyr' Permafrost Stations. Located in Yakutsk.

439 Founded from the Soil Science Department of the Biological Institute of the USSR Academy of Sciences' Siberian Division (415).

440 Founded in 1957 from the Technical Physics Department of the USSR Academy of Sciences' West Siberian Branch. In 1964 incorporated into the Institute of Semiconductor Physics (442).

441 In 1964 incorporated into the Institute of Semiconductor Physics (442).

443 In 1955 reorganized from the USSR Academy of Sciences' Sakhalin Branch. In 1957 incorporated into the USSR Academy of Sciences' Siberian Division.

444 In 1962 the Cape Schmidt Ionospheric Station of the USSR Ministry of Merchant Marine's Main North Sea Waterway was incorporated into the institute.

445 Founded in 1960 from the Irkutsk Magnetic and Ionospheric Station and the Irkutsk Region Radio Forecast Bureau.

450 Founded in 1956. In 1957 incorporated into the USSR Academy of Sciences' Siberian Division.

451 In 1964 renamed the Institute of Physical and Chemical Principles of Processing Mineral Raw Materials (452).

454 Founded in 1965 from the biological laboratories of the Institute of Experimental Biology and Medicine (453).

455 Founded in 1968 from several laboratories of the Far Eastern Department of the USSR Academy of Sciences' Siberian Division.

458	Founded in 1958. In 1967 incorporated into the Institute of the Economics and Organization of Industrial Production of the USSR Academy of Sciences' Siberian Division (460).
459	In 1958 renamed the Institute of the Economics and Organization of Industrial Production (460).
463	The Library was founded in 1918 for industrial officials. In 1958 reorganized into a public library to serve specialists in various fields of science, but mainly associates of the USSR Academy of Sciences' Siberian Division.
466	Founded in 1959. The Committee established Regional Commissions in Krasnoyarsk, Irkutsk, Yakutsk, Ulan-Ude, Vladivostok, Yuzhno-Sakhalinsk and Petropavlovsk (Kamchatka). Commission heads are the Presidium chairmen of the East Siberian, Yakutian and Far Eastern branches and the directors of comprehensive research institutes of the USSR Academy of Sciences' Siberian Division.
468	Founded from the Buryat Comprehensive Research Institute. The Buryat Institute of Natural Sciences and the Buryat Institute of Social Science were also incorporated into this branch.
469	Founded in June 1949. Until 1957 administered directly by the USSR Academy of Sciences. In 1957 transferred to the Siberian Division.
470	The V. L. Komarov Branch of the USSR Academy of Sciences was founded in 1932 and until 1957 administered directly by the USSR Academy of Sciences. In 1957 transferred to the Siberian Division.
471	Founded in 1943 and until 1957 administered directly by the USSR Academy of Sciences. In 1957 transferred to the Siberian Division. Abolished in 1959, its institutes were incorporated into the USSR Academy of Sciences' Siberian Division.

472	Founded in 1949 from the USSR Academy of Sciences' Yakutian Research Base. Until 1957 administered directly by the USSR Academy of Sciences. In 1957 transferred to the Siberian Division.

Publications of the USSR Academy of Sciences (journals)

	Title	Founded in	Publishing institution	Chief editor
1	*Avtomatika i telemekhanika* (Automation and Telemechanics)	1936	Inst of Automation and Telemechanics, USSR Acad of Sci	Trapeznikov, V.A.
2	*Avtometriya* (Autometry)	1965	Inst of Automation and Electrometry, USSR Acad of Sci's Siberian Division	Nesterikhin, Yu.Ye.
3	*Agrokhimiya* (Agrochemistry)	1964	Dept of Biochemistry, Biophysics and Active Physiological Compounds Chemistry	Rakitin, Yu.V.
4	*Aziya i Afrika segodnya* (Asia and Africa Today)	1961	Inst of Asian Peoples, USSR Acad of Sci	Gafurov, B.G.
5	*Akusticheskiy zhurnal* (Acoustics Journal)	1955	Dept of Gen and Applied Physics, USSR Acad of Sci	Grigor'yev, V.S.
6	*Astronomicheskiy vestnik* (Astronomical Herald)	1966	USSR Acad of Sci	Fedynskiy, V.V.
7	*Astronomicheskiy zhurnal* (Astronomical Journal)	1924	Dept of Gen and Applied Physics, USSR Acad of Sci	Mustel', E.R.
8	*Atomnaya energiya* (Atomic Energy)	1956	State Comt for the Use of Atomic Energy, USSR Acad of Sci	Millionshchikov, M.D.

9	*Biofizika* (Biophysics)	1956	Dept of Biochemistry, Biophysics and Active Physiological Compounds Chemistry	Rakitin, Yu.V.
10	*Biokhimiya* (Biochemistry)	1936	USSR Acad of Sci	Engel'-gardt, V.A.
11	*Botanicheskiy zhurnal* (Botanical Journal)	1916	All-Union Botanical Soc, USSR Acad of Sci	Kuprevich, V.F.
12	*Vestnik AN.SSSR* (Herald of the USSR Academy of Sciences)	1931	USSR Acad of Sci	Millionshchikov, M.D.
13	*Vestnik drevney istorii* (Herald of Ancient History)	1937	Inst of History, USSR Acad of Sci	Utchenko, S.L.
14	*Voprosy istorii* (Problems of History)	1945	Dept of History, USSR Acad of Sci	Trukhanovskiy, V.G.
15	*Voprosy ikhtiologii* (Problems of Ichthyology)	1953	Dept of Gen Biology, USSR Acad of Sci	Nikol'-skiy, G.V.
16	*Voprosy literatury* (Problems of Literature)	1957	Gorky Inst of World Lit, USSR Acad of Sci. USSR Writers' Union	Ozerov, V.
17	*Voprosy filosofii* (Problems of Philosophy)	1947	Inst of Philosophy, USSR Acad of Sci	Frolov, I.T. (1968)
18	*Voprosy ekonomiki* (Problems of Economics)	1948	Inst of Economics, USSR Acad of Sci	Khachaturov, T.S.

19	*Voprosy yazykoznaniya* (Problems of Linguistics)	1952	Inst of Linguistics, USSR Acad of Sci	Vinogradov, V.V.
20	*Vysokomolekulyarnyye soyedineniya* (High-Molecular Compounds)	1959	Dept of Gen and Tech Chemistry, USSR Acad of Sci	Medvedev, S.S. (1970)
21	*Genetika* (Genetics)	1965	Chemotechnological and Biological Section	Zhukovskiy, P.M.
22	*Geologiya i geofizika* (Geology and Geophysics)	1960	Inst of Automation and Electrometry, USSR Acad of Sci's Siberian Division	Trofimuk, A.A.
23	*Geologiya rudnykh mestorozhdeniy* (Geology of Mineral Deposits)	1959	Earth Sci Section, USSR Acad of Sci	Smirnov, V.I. (1969)
24	*Geomagnetizm i aeronomiya* (Geomagnetism and Aeronomy)	1961	Inst of Gen and Applied Physics, USSR Acad of Sci	Kalinin, Yu.D.
25	*Geotektonika* (Gcotectonics)	1965	Earth Sci Section, USSR Acad of Sci	Muratov, M.V.
26	*Geokhimiya* (Geochemistry)	1956	Earth Sci Section, USSR Acad of Sci	Vinogradov, A.P.
27	*Defektoskopiya* (Defectoscopy)	1965	Earth Sci Section, USSR Acad of Sci	Mikheyev, M.N.
28	*Doklady AN SSSR* (Reports of the USSR Academy of Sciences)	1933	Presidium, USSR Acad of Sci	Oparin, A.I.

29	*Zhurnal analiticheskoy khimii* (Journal of Analytical Chemistry)	1946	Dept of Physicochemistry and Technol of Inorganic Materials, USSR Acad of Sci	Alimarin, I.P.
30	*Zhurnal vysshey nervnoy deyatel'nosti im. I.P. Pavlova* (I.P. Pavlov Journal of Higher Nervous Activity)	1951	Dept of Physiology, USSR Acad of Sci	Asratyan, E.A.
31	*Zhurnal vychislitel'noy matematiki i matematicheskoy fiziki* (Journal of Computer Mathematics and Mathematical Physics)	1961	Dept of Mathematics, USSR Acad of Sci	Dorodnitsyn, A.A.
32	*Zhurnal nauchnoy i prikladnoy fotografii i kinematografii* (Journal of Scientific and Applied Photography and Cinematography)	1964	Chemotechnological and Biological Sci Section, USSR Acad of Sci	Chibisov, K.V.
33	*Zhurnal neorganicheskoy khimii* (Journal of Inorganic Chemistry)	1956	Dept of Gen and Tech Chemistry, USSR Acad of Sci	Spitsyn, V.I. (1967)
34	*Zhurnal obshchey biologii* (Journal of General Biology)	1940	Dept of Gen Biology, USSR Acad of Sci	Bykhovskiy, B.Ye.
35	*Zhurnal obshchey khimii* (Journal of General Chemistry)	1931	Dept of Gen and Tech Chemistry, USSR Acad of Sci	Danilov, S.N.

36	*Zhurnal organicheskoy khimii* (Journal of Organic Chemistry)	1965	Dept of Gen and Tech Chemistry, USSR Acad of Sci	Petrov, A.A.
37	*Zhurnal prikladnoy mekhaniki i tekhnicheskoy fiziki* (Journal of Applied Mechanics and Technical Physics)	1960	Siberian Division, USSR Acad of Sci	Ovsyannikov, V.V.
38	*Zhurnal prikladnoy khimii* (Journal of Applied Chemistry)	1928	Dept of Gen and Tech Chemistry, USSR Acad of Sci	Nikitin, N.I.
39	*Zhurnal strukturnoy khimii* (Journal of Structural Chemistry)	1960	Siberian Division, USSR Acad of Sci	Voyevodskiy, V.V.
40	*Zhurnal tekhnicheskoy fiziki* (Journal of Technical Physics)	1931	Dept of Gen and Applied Physics, USSR Acad of Sci	Konstantnov, B.P.
41	*Zhurnal fizicheskoy khimii* (Journal of Physical Chemistry)	1930	Dept of Gen and Tech Chemistry, USSR Acad of Sci	Gerasimov, Ya.I.
42	*Zhurnal evolyutsionnoy biokhimii i fiziologii* (Journal of Evolutionary Biochemistry and Physiology)	1965	Dept of Physiology, USSR Acad of Sci	Kreps, Ye.M.
43	*Zhurnal eksperimental'noy i teoreticheskoy fiziki* (Journal	1930	Physicotechnical and Mathematical Sci Section, USSR	Kapitsa, P.L.

	of Experimental and Theoretical Physics)		Acad of Sci	
44	*Zapiski Vsesoyuznogo mineralogicheskogo obshchestva* (Notes of the All-Union Mineralogical Society)	1866	Earth Sci Section, USSR Acad of Sci	Tatarinov, P.M.
45	*Zashchita metallov* (Metal Protection)	1965	Chemotechnological and Biological Sci Section, USSR Acad of Sci	Kolotyrkin, Ya.M.
46	*Zemlya i Vselennaya* (Earth and Universe)	1965	Physicotechnical and Mathematical Sci Section, USSR Acad of Sci	Avsyuk, G.A.
47	*Zoologicheskiy zhurnal* (Zoological Journal)	1916	Dept of Gen Biology, USSR Acad of Sci	Pavlovskiy, Ye.N. (1960)
48	*Izvestiya AN SSSR. Neorganicheskiye materialy* (News of the USSR Academy of Sciences. Inorganic Materials)	1965	Dept of Physicochemistry and Technol of Inorganic Materials, USSR Acad of Sci	Tananayev, I.V.
49	*Izvestiya AN SSSR. Mekhanika tvyordogo tela* (News of the USSR Academy of Sciences. Solid-State Mechanics)	1969	Dept of Mechanics and Control Processes, USSR Acad of Sci	

50	*Izvestiya AN SSSR. Seriya biologicheskaya* (News of the USSR Academy of Sciences. Biological Series)	1936	Chemotechnological and Biological Sci Section, USSR Acad of Sci	Mishustin, Ye.N. (1966)
51	*Izvestiya AN SSSR. Seriya geograficheskaya* (News of the USSR Academy of Sciences. Geographical Series)	1937	Dept of Earth Sci, USSR Acad of Sci	Gerasimov, I.P.
52	*Izvestiya AN SSSR. Seriya geologicheskaya* (News of the USSR Academy of Sciences. Geological Series)	1936	Dept of Earth Sci, USSR Acad of Sci	Vinogradov, A.P.
53	*Izvestiya AN SSSR. Seriya literatury i yazyka* (News of the USSR Academy of Sciences. Literature and Language Series)	1940	Dept of Lit and Language, USSR Acad of Sci	Blagoy, D.D.
54	*Izvestiya AN SSSR. Seriya matematicheskaya* (News of the USSR Academy of Sciences. Mathematical Series)	1937	Dept of Mathematics, USSR Acad of Sci	Vinogradov, I.M.
55	*Izvestiya AN SSSR. Seriya "Metally"* (News of the USSR Academy of Sciences. Metals Series)	1965	Dept of Physicochemistry and Technol of Inorganic Materials, USSR Acad of Sci	Ageyev, N.V.

56	*Izvestiya AN SSSR. Seriya "Mekhanika zhidkosti i gaza"* (News of the USSR Academy of Sciences. Liquid and Gas Mechanics Series)	1965	Dept of Mechanics and Control Processes, USSR Acad of Sci	Petrov, G.I
57	*Izvestiya AN SSSR. Seriya "Tekhnicheskaya kibernetika"* (News of the USSR Academy of Sciences. Technical Cybernetics Series)	1964	Dept of Mechanics and Control Processes, USSR Acad of Sci	Petrov, B.N.
58	*Izvestiya AN SSSR. Fizika atmosfery i okeana* (News of the USSR Academy of Sciences. Atmospheric and Oceanic Physics Series)	1965	Dept of Earth Sci, USSR Acad of Sci	Obukhov, A.M.
59	*Izvestiya AN SSSR. "Fizika Zemli"* (News of the USSR Academy of Sciences. Earth Physics)	1965	Dept of Earth Sci, USSR Acad of Sci	Riznichenko, Yu.V.
60	*Izvestiya AN SSSR. Seriya khimicheskaya* (News of the USSR Academy of Sciences. Chemical Series)	1936	Chemotechnological and Biological Sci Section, USSR Acad of Sci	Dubinin, M.M.

61	*Izvestiya AN SSSR. Energetika i transport* (News of the USSR Academy of Sciences. Power Engineering and Transport)	1963	Dept of Physico-tech Power Eng Problems, USSR Acad of Sci	Popkov, V.I.
62	*Izvestiya Vsesoyuznogo geograficheskogo obshchestva* (News of the All-Union Geographical Society)	1865	Geographical Soc, USSR Acad of Sci	Kolesnik, S.V.
63	*Izvestiya SO AN SSSR* (News of the USSR Academy of Sciences' Siberian Division)	1965	USSR Acad of Sci	Struminskiy, V.V. (1968)
63a	*Seriya biologomeditsinskikh nauk* (Biomedical Sciences Series)	1965	Siberian Division, USSR Acad of Sci	Krylov, G.V.
63b	*Seriya obshchestvennykh nauk* (Social Sciences Series)	1965	Siberian Division, USSR Acad of Sci	Prudenskiy, G.A.
63c	*Seriya tekhnicheskikh nauk* (Technical Sciences Series)	1965	Siberian Division, USSR Acad of Sci	Zhukov, M.F.
63d	*Seriya khimicheskikh nauk* (Chemical Series)	1965	Siberian Division, USSR Acad of Sci	Nikolayev, A.V.
64	*Istoriya SSSR* (History of the USSR)	1957	Dept of History, USSR Acad of Sci	Koval'-chenko, I.D. (1969)

65	*Kinetika i kataliz* (Kinetics and Catalysis)	1960	Siberian Division, USSR Acad of Sci	Boreskov, G.K.
66	*Kolloidnyy zhurnal* (Colloidal Journal)	1935	Dept of Gen and Tech Chemistry, USSR Acad of Sci	Rebinder, P.A.(1968)
67	*Kosmicheskiye issledovaniya* (Space Research)	1963	Presidium, USSR Acad of Sci	Sedov, L.I.
68	*Kristallografiya* (Crystallography)	1956	Dept of Gen and Applied Physics, USSR Acad of Sci	Belov, N.B.(1968)
69	*Litologiya i poleznyye iskopayemyye* (Lithology and Minerals)	1963	Dept of Earth Sci, USSR Acad of Sci	Bushinskiy, I.G.(1969)
70	*Matematicheskiy sbornik* (Mathematical Collection)	1866	Dept of Mathematics, USSR Acad of Sci. Moscow Mathematical Soc	Petrovskiy, I.G.
71	*Mashinovedeniye* (Machine Science)	1965	Dept of Mechanics and Control Processes, USSR Acad of Sci	Blagonravov, A.A.
72	*Latinskaya Amerika* (Latin America)	?	USSR Acad of Sci	Mikoyan, S.A.
73	*Lesovedeniye* (Forestry Management)	1967	USSR Acad of Sci	Sukachyov, V.N.
74	*Matematicheskiye zametki* (Mathematical Notes)	1966	USSR Acad of Sci	Stechkin, S.B.

75	*Mikologiya i fitopatologiya* (Mycology and Phytopathology)	1967	USSR Acad of Sci	Gorlenko, M.V.
76	*Mikrobiologiya* (Microbiology)	1932	Dept of Biochemistry, Biophysics and Active Physiological Compounds Chemistry	Imshenetskiy, A.A.
77	*Mirovaya ekonomika i mezhdunarodnyye otnosheniya* (World Economics and International Relations)	1957	Inst of World Economics and International Relations, USSR Acad of Sci	Khavinson, Ya.S.
78	*Narody Azii i Afriki* (Afro-Asian Peoples)	1961	Inst of Asian Peoples, USSR Acad of Sci	Braginskiy, I.S.
79	*Nauchnotekhnicheskaya informatsiya* (Scientific and Technical Information)	1963	Inst of Sci and Tech Information, USSR Acad of Sci	Mikhaylov, A.I.
80	*Neftekhimiya* (Petrochemistry)	1961	Dept of Gen and Tech Chemistry, USSR Acad of Sci	Rebinder, P.A. (1968)
81	*Novaya i noveyshaya istoriya* (Modern and Recent History)	1957	Inst of History, USSR Acad of Sci	Narochnitskiy, A.L.
82	*Okeanologiya* (Oceanology)	1961	Oceanographic Commission, USSR Acad of Sci	Zenkevich, L.A.
83	*Ontogenez* (Ontogenesis)	1969	USSR Acad of Sci	Astaurov, B.L. (1969)

84	*Optika i spektroskopiya* (Optics and Spectroscopy)	1956	Dept of Gen and Applied Physics, USSR Acad of Sci	Frish, S.E.
85	*Paleontologicheskiy zhurnal* (Paleontological Journal)	1959	Dept of Gen Biology, USSR Acad of Sci	Ruzhentsev, V.Ye. (1968)
86	*Parazitologiya* (Parasitology)	1967	USSR Acad of Sci	Bykhovskiy, B.Ye.
87	*Pochvovedeniye* (Soil Science)	1899	Chemotechnological and Biological Sci Section, USSR Acad of Sci	Peyve, A.Ya.
88	*Pribory i tekhnika eksperimenta* (Experimental Instruments and Equipment)	1956	Physicotech and Mathematical Sci Section, USSR Acad of Sci	Shal'nikov, A.I.
89	*Prikladnaya biokhimiya i mikrobiologiya* (Applied Biochemistry and Microbiology)	1965	Dept of Biochemistry, Biophysics and Active Physiological Compounds Chemistry, USSR Acad of Sci	Bukin, V.N.
90	*Prikladnaya matematika i mekhanika* (Applied Mathematics and Mechanics)	1933	Dept of Mechanics and Control Processes, USSR Acad of Sci	Galin, L.A.
91	*Priroda* (Nature)	1912	Presidium, USSR Acad of Sci	Basov, N.G.(1967)
92	*Problemy peredachi informatsii* (Data Transmission Problems)	1965	Dept of Mechanics and Control Processes, USSR Acad of Sci	Siforov, V.I. (1966)

93	*Radiobiologiya* (Radiobiology)	1961	Dept of Biochemistry, Biophysics and Active Physiological Compounds Chemistry, USSR Acad of Sci	Kuzin, A.M.
94	*Radiotekhnika i elektronika* (Radio Engineering and Electronics)	1956	Dept of Gen and Applied Physics, USSR Acad of Sci	Kotel'nikov, V.A.
95	*Radiokhimiya* (Radiochemistry)	1959	Dept of Gen and Tech Chemistry, USSR Acad of Sci	Vdovenko, V.M.
96	*Rastitel'nyye resursy* (Vegetative Resources)	1965	Dept of Gen Biology, USSR Acad of Sci	Fyodorov, A.A.
97	*Referativnyy zhurnal* (Abstracts Journal)	1953	Presidium, USSR Acad of Sci	Mikhaylov, A.I.
98	*Russkaya literatura* (Russian Literature)	1958	Inst of Russian Lit, USSR Acad of Sci	Bazanov, V.G.
99	*Russkaya rech'* (Russian Speech)	1966	USSR Acad of Sci	Borkovskiy, V.I. (1967)
100	*Sibirskiy matematicheskiy zhurnal* (Siberian Mathematical Journal)	1960	Siberian Division, USSR Acad of Sci	Sobolev, S.L. (1968)
101	*Sovetskaya arkheologiya* (Soviet Archeology)	1957	Inst of Archeology, USSR Acad of Sci	Artsikhovskiy, A.V.

102	*Sovetskaya tyurkologiya* (Soviet Turcology)	1968	USSR Acad of Sci	Shiraliyev, M.Sh.
103	*Sovetskaya etnografiya* (Soviet Ethnography)	1931	Inst of Ethnography, USSR Acad of Sci	Petrov-Averkiyev, Yu.P. (1966)
104	*Sovetskiye arkhivy* (Soviet Archives)	1966	Main Archives Bd (publ by Inst of History, USSR Acad of Sci and Inst of Marxism-Leninism)	Kondrat'yev, V.A.
105	*Sovetskoye gosudarstvo i pravo* (Soviet State and Law)	1927	Inst of State and Law, USSR Acad of Sci	Ivanov, S.A.
106	*Sovetskoye slavyanovedeniye* (Soviet Slavic Studies)	1965	Inst of Slavic Studies, USSR Acad of Sci	Kostyushko, I.I.
106a	*SShA - ekonomika, politika, ideologiya* (USA - Economics, Politics and Ideology)	1968	USSR Acad of Sci	Berezhkov, V.M. (1969)
107	*Teoriya veroyatnostey i yeyo primeneniye* (Probability Theory and Its Application)	1956	Dept of Mathematics, USSR Acad of Sci	Prokhorov, Yu.V. (1967)
108	*Teoreticheskaya i matematicheskaya fizika* (Theoretical and Mathematical Physics)	?	USSR Acad of Sci	Bogolyubov, N.N. (1969)

109	*Teoreticheskiye osnovy khimicheskoy tekhnologii* (Theoretical Principles of Chemical Technology)	1967	USSR Acad of Sci	Zhavoronkov, N.M.
110	*Teplofizika vysokikh temperatur* (High-Temperature Thermal Physics)	1963	Dept of Physicotech Problems of Power Eng, USSR Acad of Sci	Sheyndlin, A.Ye.
111	*Teploenergetika* (Thermal Power Engineering)	1954	State Comt for Sci and Tech, USSR Council of Min. USSR Acad of Sci	Vukalovich, M.P.
112	*Uspekhi matematicheskikh nauk* (Progress of Mathematical Sciences)	1936	Dept of Mathematics, USSR Acad of Sci	Aleksandrov, P.S.
113	*Uspekhi sovremennoy biologii* (Progress of Modern Biology)	1932	Chemotechnological and Biological Sci Section, USSR Acad of Sci	Belozerskiy, A.N.
114	*Uspekhi fiziologicheskikh nauk* (Progress of Physiological Sciences)	?	USSR Acad of Sci	Parin, V.V. (1969)
115	*Uspekhi fizicheskikh nauk* (Progress of Physical Sciences)	1918	Physicotech and Mathematical Sci Section, USSR Acad of Sci	Shpol'skiy, E.V.
116	*Uspekhi khimii* (Progress of Chemistry)	1932	Chemotechnological and Biological Sci Section, USSR Acad of Sci	Korshak, V.V.

117	*Fizika goreniya i vzryva* (Combustion and Explosion Physics)	1966	Siberian Division, USSR Acad of Sci	Lavrent'yev, M.A.
118	*Fizika i tekhnika poluprovodnikov* (Semiconductor Physics and Engineering)	?	USSR Acad of Sci	Ryvkin, S.M. (1967)
119	*Fizika i khimiya obrabotki materialov* (Material Processing Physics and Chemistry)	1967	USSR Acad of Sci	Rykalin, N.N.
120	*Fizika metallov i metallovedeniye* (Metal Physics and Metal Science)	1955	Dept of Gen and Applied Physics, USSR Acad of Sci	Vonsovskiy, S.V.
121	*Fizika tvyordogo tela* (Solid-State Physics)	1959	Dept of Gen and Applied Physics, USSR Acad of Sci	Zhurkov, S.N.
122	*Fiziko-tekhnicheskiye problemy razrabotki poleznykh iskopayemykh* (Physicotechnical Problems of Working Minerals)	1966	Siberian Division, USSR Acad of Sci	Gorbachyov, T.F.
123	*Fiziologicheskiy zhurnal im. I.M. Sechenova* (I.M. Sechenov Physiological Journal)	1917	I.P. Pavlov All-Union Physiological Soc, USSR Acad of Sci	Chernigovskiy, V.N. (1969)

124	*Fiziologiya rasteniy* (Physiology of Plants)	1954	Dept of Biochemistry, Biophysics and Active Physiological Compounds Chemistry, USSR Acad of Sci	Kursanov, A.L.
125	*Funktsional'- nyy analiz i yego prilozheniya* (Functional Analysis and Its Application)	1966	USSR Acad of Sci	Gel'fand, I.M.
126	*Khimiya vysokikh energiy* (High-Energy Chemistry)	1967	USSR Acad of Sci	Gol'danskiy, V.I.
127	*Khimiya i zhizn'* (Chemistry and Life)	1965	Chemotechnological and Biological Sci Section, USSR Acad of Sci	Petryakov-Sokolov, I.V.
128	*Khimiya i tekhnologiya topliv i masel* (Fuels and Oils Chemistry and Technology)	1956	Chemotechnological and Biological Sci Section, USSR Acad of Sci	Polyakov, I.S.
129	*Khimiya tvyordogo topliva* (Solid Fuel Chemistry)	1967	USSR Acad of Sci	Karavayev, N.M.
130	*Tsitologiya* (Cytology)	1959	Dept of Biochemistry, Biophysics and Active Physiological Compounds Chemistry, USSR Acad of Sci	Troshin, A.S.
131	*Ekologiya* (Ecology)	?	USSR Acad of Sci	Shvarts, S.S. (1969)

132	*Ekonomika i matematiches- kiye metody* (Economics and Mathematical Methods)	1964	Centr Economi- cal Mathematics Inst, USSR Acad of Sci	Fedorenko, N.P.
133	*Elektrichestvo* (Electricity)	1880	Dept of Physico- tech Problems of Power Eng, USSR Acad of Sci	Razevig, D.V.
134	*Elektrokhimiya* (Electrochem- istry)	1965	Dept of Gen and Tech Chemistry, USSR Acad of Sci	Frumkin, A.N.
135	*Entomologiches- koye obozreniye* (Entomological Review)	1956	All-Union Ento- mological Soc, USSR Acad of Sci	Shtakel'- berg, A.A.
136	*Yadernaya fi- zika* (Nuclear Physics)	1965	Dept of Nuclear Physics, USSR Acad of Sci	Veksler, V.I.

Chapter III

MEMBERS OF THE USSR ACADEMY OF SCIENCES

Full, Corresponding and Honorary Members of the USSR Academy of Sciences

Name	Date of Birth Death	Profession	Corresp Member from	Full	Hon
Abdullayev, Kh.M.	1912-1962	geologist	1958		
Abramovich, D.I.	1873-1955	lit historian	1921		
Abrikosov, A.A.	1928-	physicist	1964		
Abrikosov, A.I.	1875-1955	pathoanatomist		1939	
Adoratskiy, V.V.	1878-1945	philosopher		1932	
Adrianova-Perets, V.P.	1888-	historian	1943		
Afanas'yev, G.D.	1906-	geologist	1953		
Aganbegyan, A.G.	1932-	economist	1964		
Ageyev, N.V.	1903-	chemist, metallurgist	1946	1968	
Agoshkov, M.I.	1905-	mining eng	1953		
Akhvlediani, G.S.	1887-	linguist	1939		

Name	Dates	Field		
Akimov, G.V.	1901-1953	physico-chemist	1939	
Aksenyonok, G.A.	1910-	lawyer	1966	
Alekin, O.A.	1908-	chemist	1953	
Aleksandrov, A.D.	1912-	mathematician	1946	1964
Aleksandrov, A.P.	1903-	physicist	1943	1953
Aleksandrov, B.K.	1889-	hydrotechnician	1953	
Aleksandrov, G.F.	1908-1961	philosopher		1946
Aleksandrov, I.G.	1875-1936	power eng		1932
Aleksandrov, P.S.	1896-	mathematician	1929	1953
Alekseyev, A.Ye.	1891-	electr eng	1953	
Alekseyev, M.P.	1896-	lit historian	1946	1958
Alekseyev, V.M.	1881-1951	sinologist		1929
Alekseyevskiy, N.Ye.	1912-	physicist	1960	
Alikhanov, A.I.	1904-1970	physicist	1939	1943
Alikhan'yan, A.I.	1908-	physicist	1946	
Alimarin, I.P.	1903-	chemist	1953	1966
Ambartsumyan, R.S.	1911-	mech eng	1966	
Ambartsumyan, V.A.	1908-	astrophysicist	1939	1953

Amiranashvili, Sh.Ya.	1899-	historian	1943	
Amiraslanov, A.A.-ogly	1900-1962	geologist	1953	
Andreyev, M.S.	1873-1948	ethnographer	1929	
Andreyev, N.N.	1880-	physicist	1933	1953
Andrianov, K.A.	1904-	chemist	1953	1964
Andriashev, A.P.	1910-	zoologist	1966	
Andronov, A.A.	1901-1952	physicist		1946
Anichkov, N.N.	1885-1964	pathomorphologist		1939
Anisimov, I.I.	1899-1966	lit historian	1960	
Anozin, P.K.	1898-	physiologist		1966
Antipin, P.F.	1890-1960	metallurgist	1939	
Anuchin, D.N.	1843-1923	anthropologist, ethnographer	1896	1898
Arbuzov, A.Ye.	1877-1968	chemist	1932	1942
Arbuzov, B.A.	1903-	chemist	1943	1953
Arkad'yev, V.K.	1884-1953	physicist	1927	
Arkhangel'skiy, A.D.	1879-1940	geologist	1925	1929
Arkhangel'skiy, A.S.	1854-1926	historian	1904	
Arkhangel'skiy, S.I.	1882-1958	historian	1946	

Artobolevskiy, I.I.	1905-	mech eng	1939	1946
Artsikhovskiy, A.V.	1902-	archeologist	1960	
Artsimovich, L.A.	1909-	physicist	1946	1953
Arzhanov, M.A.	1902-1960	lawyer	1939	
Arzumanyan, A.A.	1904-1965	economist	1958	1962
Asaf'yev, B.V.	1884-1949	musicologist		1943
Asratyan, E.A.	1903	physiologist	1939	
Astaurov, B.L.	1904-	biologist	1958	1966
Avakyan, A.A.	1907-1966	biologist	1946	
Avanesov, R.I.	1902-	linguist	1958	
Avdeyev, V.N.	1915-	electronics specialist	1958	
Averbakh, M.I.	1872-1944	ophthalmologist		1939
Avrorin, V.A.	1907-	linguist	1964	
Avsyuk, G.A.	1906-	glacialist	1960	
Babayev, Yu.N.	1928-	physicist	1968	
Bagdasar'yan, Kh.S.	1908-	physicochemist	1968	
Bakh, A.N.	1857-1946	biochemist		1929
Bakhrakh, L.D.	1921-	radiophysicist	1966	

Bakhrushin, S.V.	1882-1950	historian	1939	
Bakhurin, I.M.	1880-1940	geophysicist	1939	
Bakulev, A.N.	1890-1967	surgeon		1958
Balandin, A.A.	1898-1967	organic chemist	1943	1946
Barannikov, A.P.	1890-1952	indologist		1939
Baranov, P.A. *	1892-1962	botanist	?	
Baranskiy, N.N.	1881-1963	geographer	1939	
Bardin, I.P.	1883-1960	metallurgist		1932
Barkhudarov, S.G.	1894-	linguist	1946	
Barmin, V.P.	1909-	mech eng	1958	1966
Bartol'd, V.V.	1869-1930	Orientalist	1910	1912
Bashkirov, A.N.	1903-	organic chemist	1958	
Basov, N.G.	1922	physicist	1962	1966
Bayev, A.A.	1903-	biochemist	1968	
Baykov, A.A.	1870-1946	metals specialist		1932
Bazanov, V.G.	1911-	historian	1962	
Bel'chikov, N.F.	1890-	lit historian	1953	

---

* 1948 was already corresp member, USSR Acad of Sci.

Beletskiy, A.I.	1884-1961	lit historian	1946	1958
Belopol'skiy, A.A.	1854-1934	astronomer	1903	1906
Belousov, V.V.	1907-	geologist	1953	
Belov, N.V.	1891-	crystallographer	1946	1953
Belozerskiy, A.N.	1905-	biochemist	1958	1962
Belyayev, A.I.	1906-1967	metals specialist	1960	
Belyayev, D.K.	1917-	zoologist	1964	
Belyayev, N.M.	1890-1944	mech eng	1939	
Belyayev, S.T.	1923-	physicist	1964	1968
Belyankin, D.S.	1876-1953	geologist	1933	1943
Belyavskiy, S.I.	1883-1953	astronomer	1939	
Beneshevich, V.N.	1874-1943	Byzantologist	1924	
Berg, A.I.	1893-	radio eng	1943	1946
Berg, L.S.	1876-1950	geographer, zoologist	1928	1946
Bergel'son, L.D.	1918-	chemist	1968	
Beritashvili, I.S.	1884-	physiologist		1939
Berkov, P.N.	1896-1969	lit historian	1960	
Bernshteyn, S.N.	1880-1968	mathematician	1924	1929

Bertel's, Ye.E.	1890–1957	Orientalist	1939	
Betekhtin, A.G.	1897–1962	mineralogist	1946	1953
Bey-Biyenko, G.Ya.	1903–	entomologist	1953	
Bezrukov, P.L.	1909–	oceanologist	1968	
Bilibin, Yu.A.	1901–1952	geologist	1946	
Bitsadze, A.V.	1916–	mathematician	1958	
Blagonravov, A.A.	1894–	mech eng		1943
Blagoy, D.D.	1893–	lit historian	1953	
Blazhko, S.N.	1870–1956	astronomer	1929	
Blinova, Ye.N.	1906–	meteorologist	1953	
Blokhintsev, D.I.	1908–	physicist	1958	
Bochvar, A.A.	1902–	metals specialist	1939	1946
Bogolepov, M.I.	1879–1945	economist	1939	
Bogolyubov, M.N.	1918–	linguist	1966	
Bogolyubov, N.N.	1909–	mathematician	1946	1953
Bogomolets, A.A.	1881–1946	pathophysiologist		1932
Bogomolov, A.F.	1913–	radiophysicist	1966	
Bogoroditskiy, V.A.	1857–1941	linguist	1915	

Bogorov, V.G.	1904-	oceanologist	1958	
Bogoslovskiy, M.M.	1867-1929	historian		1921
Bogoyavlenskiy, S.K.	1871-1947	historian	1929	
Bokiy, G.B.	1909-	chemist	1958	
Bol'shakov, K.A.	1906-	chemist	1958	
Bonch-Bruyevich, M.A.	1888-1940	radio eng	1931	
Boreskov, G.K.	1907-	physicochemist	1958	1966
Borisyak, A.A.	1872-1944	paleontologist	1923	1929
Borkovskiy, V.I.	1900-	linguist	1958	
Borodin, I.P.	1847-1930	botanist	1887	1902
Borovik-Romanov, V.-A.S.	1920-	physicist	1966	
Borovkov, A.A.	1931-	mathematician	1966	
Borovkov, A.K.	1904-1962	Turcologist	1958	
Braunshteyn, A.Ye.	1902-	biochemist	1960	1964
Brekhovskikh, L.M.	1917-	physicist	1953	1968
Brilling, N.R.	1876-1961	thermal eng	1953	
Britske, E.V.	1877-1953	metallurgist	1931	1932
Brodskiy, A.I.	1895-1969	physicochemist	1943	

Bromley, Yu.V.	1921–	historian	1966	
Bruk, I.S.	1902–	electr eng	1939	
Bruyevich, N.G.	1896–	mech eng	1939	1942
Bubrikh, D.V.	1890–1949	linguist	1946	
Budker, G.I.	1918–	physicist	1958	1964
Budnikov, P.P.	1885–1968	chemist	1939	
Budyko, M.I.	1920–	geophysicist	1964	
Bukharin, N.I.*	1888–1938	economist		?
Bukin, V.N.	1899–	biochemist	1964	
Bulakhovskiy, L.A.	1888–1961	Slavist	1946	
Bulanzhe, Yu.D.	1911–	geologist, geophysicist	1966	
Bulgakov, B.V.	?	technician	1946	
Bunin, I.A.	1870–1953	writer		1909
Bunkin, B.V.	1922–	radio eng	1968	
Burdenko, N.N.	1876–1946	surgeon		1939
Bush, N.A.	1869–1941	botanist	1920	
Bushinskiy, V.P.	1885–1960	soil scientist	1939	
Bushmin, A.S.	1910–	lit historian	1960	

* In May 1937 the Gen Assembly of the USSR Acad of Sci expelled Bukharin from the Academy under Paragraph 24 of the Academy's Statutes.

Bushuyev, K.D.	1914-	mech eng	1960	
Buslayev, Yu.A.	1929-	inorganic chemist	1968	
Buslenko, N.P.	1922-	mathematician	1966	
Butkevich, V.S.	1872-1942	plant physiologist, biochemist	1929	
Buzyskul, V.P.	1858-1931	historian		1922
Bychkov, I.A.	1858-1944	archeographer	1903	
Bykhovskiy, B.Ye.	1908-	parasitologist	1960	1964
Bykov, K.M.	1886-1959	physiologist		1946
Byushgens, G.S.	1916-	mech eng	1966	
Chagin, B.A.	1899-	philosopher	1960	
Chaplygin, S.A.	1869-1942	hydraulic and aviation eng	1924	1929
Chaylakhyan, M.Kh.	1902-	physiologist	1968	
Chebotarev, N.G.	1894-1947	mathematician	1929	
Chekmarev, A.P.	1902-	metallurgist		1968
Chelintsev, V.V.	1877-1947	organic chemist	1933	
Chelomey, V.N.	1914-	mech eng	1958	1962
Chepikov, K.R.	1900-	geologist	1953	
Cherenkov, P.A.	1904-	physicist	1964	
Chernigovskiy, V.N.	1907-	physiologist	1953	1960

Chernyshyov, A.A.	1882-1940	electr eng	1929	1932
Chernyshyov, A.B.	1904-1953	gasification eng	1939	
Chernyshyov, V.I.	?	?	1939	
Chyornyy, G.G.	1923-	mech eng	1962	
Chernyayev, I.I.	1893-1966	inorganic chemist	1933	1943
Cherskiy, N.V.	1905-	mech eng	1968	
Chertok, B.Ye.	1912-	cybernetics specialist	1968	
Chetayev, N.G.	1902-1959	mech eng	1943	
Chibisov, K.V.	1897-	photography specialist	1946	
Chichibabin, A.Ye.*	1871-1945	inorganic chemist	1926	1928-1936
Chinakal, N.A.	1888-	mining eng	1958	
Chizhevskiy, N.P.	1873-1952	metallurgist		1939
Chizhikov, D.M.	1895-	metallurgist	1939	
Chkhikvadze, V.M.	1912-	lawyer	1964	
Chmutov, K.V.	1902-	physico-chemist	1953	
Chudakov, A.Ye.	1921-	physicist	1966	
Chudakov, Ye.A.	1890-1953	mech eng	1933	1939

---

* In 1930 emigrated. In 1936 expelled from the Academy.

Chufarov, G.I.	1900-	physico-chemist	1953	
Chukhanov, Z.F.	1912-	thermal and power eng	1939	
Chukhrov, F.V.	1908-	geochemist	1953	
Danilov, S.N.	1889-	chemist	1943	
Deborin, A.M.	1881-1963	philosopher, historian		1929
Delone, B.N.	1890-	mathematician	1929	
Dem'yanov, N.A.	1861-1938	organic chemist	1927-1929	
Deryagin, B.V.	1902-	physico-chemist	1946	
Derzhavin, N.S.	1877-1953	Slavist		1931
Desnitskaya, A.V.	1912-	linguist	1964	
Devyatkov, N.D.	1907-	electr eng	1953	1968
Devyatykh, G.G.	1918-	inorganic chemist	1968	
Dikushin, V.I.	1902-	mech eng	1943	1953
Dinnik, A.N.	1876-1950	mech eng		1946
Dmitriyev, N.K.	1898-1954	linguist	1943	
Dobiash-Rozhdestvenskaya, O.A.	1874-1939	historian	1929	
Dobroklonskiy, M.V.	1886-1964	art critic	1943	
Dobrovol'skiy, V.V.	1880-1956	mech eng	1946	

Name	Dates	Field		
Dogel', V.A.	1882-1955	zoologist	1939	
Dolgoplosk, B.A.	1905-	chemist	1958	1964
Dollezhal', N.A.	1899-	thermal eng	1953	1962
Dorodnitsyn, A.A.	1910-	geophysicist		1953
Druzhinin, N.M.	1886-	historian		1953
Druzhinin, S.I.	?	technician	1933	
Dubinin, M.M.	1901-	physico-chemist		1943
Dubinin, N.P.	1907-	biologist	1946	1966
Dukhov, N.L.	1904-1964	mech eng	1953	
Dumanskiy, A.V.	1880-1967	chemist	1933	
D'yachenko, V.P.	1902-	economist	1953	
D'yakonov, M.A.	1855-1919	law historian	1909	1912
Dynnik, M.A.	1896-	philosopher	1958	
Dzhanashiya, S.N.	1900-1947	historian		1943
Dzhavakhash-vili, I.A.	1876-1940	historian		1939
Dzhelepov, B.S.	1910-	physicist	1953	
Dzhelepov, V.P.	1913-	physicist	1966	
Dzotsenidze, G.S.	1910-	mineralogist		1968
Emanuel', N.M.	1915-	physico-chemist	1958	1966

Endzelin, Ya. (Yan Martsevich)	1873-1961	linguist	1929	
Eneyev, T.M.	1924	mech eng	1968	
Engel'gardt, V.A.	1894-	biochemist	1946	1953
Eykhfel'd, I.G.	1893-	botanist	1953	
Fadeyev, D.K.	1907-	mathematician	1964	
Famintsyn, A.S.	1835-1918	botanist	1884	1891
Farmakovskiy, B.V.	1870-1928	archeologist	1914	
Favorskiy, A.Ye.	1860-1945	organic chemist	1921	1929
Fedchenko, O.A.	1845-1921	botanist	1906	
Fedin, K.A.	1892-	writer		1958
Fedorenko, N.P.	1917-	economist	1962	1964
Fedorenko, N.T.	1912-	lit historian	1958	
Fedoseyev, P.N.	1908-	philosopher	1946	1960
Fedot'yev, P.P.	1864-1934	chemical technol	1933	
Fedynskiy, V.V.	1908-	geophysicist	1968	
Feofilov, P.P.	1915-	physicist	1964	
Feoktistov, L.P.	1928-	physicist	1966	
Ferdman, D.L.	1903-1970	biochemist	1946	
Fersman, A.Ye.	1883-1945	mineralogist		1919
Fesenkov, V.G.	1889-	astronomer	1927	1935

Feynberg, Ye.L.	1912-	physicist	1966	
Filin, F.P.	1908-	linguist	1962	
Flavitskiy, F.M.	1848-1917	organic chemist	1907	
Flerov, G.N.	1913-	physicist	1953	1968
Florensov, N.A.	1909-	geologist	1960	
Florin, V.A.	1899-1960	soil scientist	1953	
Fok, V.A.	1898-	physicist	1932	1939
Fokin, A.V.	1912-	organic chemist	1968	
Fomin, V.V.	1909-	chemical technol	1964	
Fotiadi, E.E.	1907-	geologist, geophysicist	1958	
Frank, G.M.	1904-	biophysicist	1960	1966
Frank, I.M.	1908-	physicist	1946	1968
Frantsov, G.P.	1903-1969	philosopher	1958	1964
Frenkel', Ya.I.	1894-1952	physicist	1929	
Freydlina, R.Kh.	1906-	chemist	1958	
Freyman, A.A.	1879-1968	Orientalist	1928	
Friche, V.M.	1870-1929	lit historian		1929
Frish, S.E.	1899-	physicist	1946	
Frumkin, A.N.	1895-	physico-chemist		1932

Fyodorov, A.A.	1906-	botanist	1964	
Fyodorov, S.F.	1896-1970	geologist	1939	
Fyodorov, Ye.K.	1910-	geophysicist	1939	1960
Fyodorov, Ye.S.	1853-1919	crystallog-	1901	1919
Fyodorov, Ye.Ye.	1880-1965	climatologist	1946	
Fyodorovskiy, N.M.	1886-1956	mineralogist	1933	
Gafurov, B.G.	1908-	Orientalist	1958	1968
Galerkin, B.G.	1871-1945	mech eng	1928	1935
Galin, L.A.	1912-	mech eng	1953	
Gamaleya, N.F.	1859-1949	microbiologist	1939	1940
Gamburtsev, G.A.	1903-1955	geophysicist	1946	1953
Gamov, G.A.	?	physicist	1932	
Gan, V.Yu.	?	technician	1932	
Gaponov-Grekhov, A.V.	1926-	physicist	1964	1968
Gatovskiy, L.M.	1903-	economist	1960	
Gavrilov, M.A.	1903-	automation and telemechanics specialist	1964	
Gazenko, O.G.	1918-	physiologist	1966	
Gedroyts, K.K.	1872-1932	agrochemist	1927	1929

Gel'fand, I.M.	1913-	mathematician	1953		
Gel'fond, A.O.	1906-1968	mathematician	1939		
Gerasimov, I.P.	1905-	geographer	1946	1953	
Gerasimov, Ya.I.	1903-	physico-chemist	1953		
German, A.P.	1874-1953	mining eng		1939	
Gersevanov, N.M.	1879-1950	mech eng	1939		
Gershuni, G.V.	1905-	physiologist	1964		
Gertsen, P.A.	1871-1947	surgeon	1939		
Gessen, B.M.	?	philosopher	1933		
Gilyarov, M.S.	1912-	entomologist	1966		
Ginzburg, V.L.	1916-	physicist	1953	1966	
Glazenap, S.P.	1848-1937	astronomer	1927		1929
Glinka, K.D.	1867-1927	soil scientist	1926	1927	
Glushko, V.P.	1908-	thermal eng	1953	1958	
Glushkov, V.G.	1883-1939	hydrologist	1932		
Glushkov, V.M.	1923-	mathematician		1964	
Gol'danskiy, V.I.	1923-	physico-chemist	1962		
Golubeyev, V.V.	1884-1954	mathematician mech eng	1934		
Golubtsov, V.A.	1894-	thermal eng	1953		

Golunskiy, S.A.	1895-1962	lawyer, diplomat	1939		
Gorbachyov, T.F.	1900-	mining eng	1958		
Gorbunov, N.P.	1892-1944	?		1935	
Gordlevskiy, V.A.	1876-1956	Orientalist	1929	1946	
Gorinov, A.V.	1902-	railroad transport specialist	1939		
Gorshkov, G.S.	1921-	geochemist	1966		
Gorky, Maksim* (Peshkov, A.M.)	1868-1936	writer			1902
Gor'kov, L.P.	1929-	physicist	1966		
Gorskiy, I.I.	1893-	geologist	1943		
Goryachkin, V.P.	1868-1935	technician			1929
Got'ye, Yu.V.	1873-1943	historian	1922	1939	
Grabar', I.E.	1871-1960	painter, historian		1943	
Graftio, G.O.	1869-1949	water and power eng		1932	
Grashchenkov, N.I.	1901-1965	neurologist	1939		
Grave, D.A.	1863-1939	mathematician			1929

---

* After his election as an hon member of the Acad of Sci, Gorky was deprived of membership on orders of Tsar Nikolay II without the consent of the members who had elected him. In 1917 he was readmitted as an hon member without new elections.

Grebenshchikov, I.V.	1887-1953	inorganic chemist		1932
Grekov, B.D.	1882-1953	historian	1934	1935
Grigolyuk, E.I.	1923-	mech eng	1958	
Grigor'yev, A.A.	1883-1968	geographer		1939
Grigor'yev, I.F.	1890-?	geologist		1946
Grinberg, A.A.	1898-1966	chemist	1943	1958
Grinberg, G.A.	1900-	physicist	1946	
Gross, Ye.F.	1897-	physicist	1946	
Grossgeym, A.A.	1888-1948	botanist	1939	1946
Grosul, Ya.S.	1912-	historian	1966	
Grum-Grzhimaylo, V.Ye.	1864-1928	metallurgist	1927	
Grushevskiy, M.S.	1866-1934	historian		1929
Grushin, P.D.	1906-	aviation eng	1962	1966
Guber, A.A.	1902-	historian	1953	1966
Gubkin, I.M.	1871-1939	geologist		1929
Gudtsov, N.T.	1885-1957	metallurgist		1939
Gulevich, V.S.	1867-1933	biochemist	1928	1929
Gurevich, I.I.	1912-	physicist	1968	
Gutyrya, V.S.	1910-	chemist	1943	

Name	Dates	Field		
Gyunter, R.M.	1871-1941	mathematician	1924	
Ignatovskiy, V.S.	?	technician	1932	
Ikonnikov, V.S.	1841-1923	historian	1863	1914
Il'ichyov, A.S.	1898-1952	mining eng	1939	
Il'ichyov, A.S.	1906-	philosopher		1962
Il'inskiy, G.A.*	?	linguist	?	
Il'inskiy, M.A.	1856-1941	chemist		1935
Il'yushin, A.A.	1911-	mech eng	1943	
Il'yushin, S.V.	1894-	plane designer		1968
Imshenetskiy, A.A.	1905-	microbiologist	1946	1962
Inostrantsev, A.A.	1843-1919	geologist	1901	
Inozemtsev, N.N.	1921-	economist	1964	1968
Ioffe, A.F.	1880-1960	physicist	1918	1920
Iovchuk, M.T.	1908-	philosopher	1946	
Ipat'yev, V.N.**	1867-1952	organic chemist	1914	1916-1936
Isachenko, B.L.	1871-1948	microbiologist	1933	1946

---

* 1948 recorded as corresp member, USSR Acad of Sci.

** Emigré; Dec 1936 session of the USSR Acad of Sci stripped him of membership for refusing to return to the USSR.

Isakov, I.S.	1894-1967	oceanologist	1958	
Ishlinskiy, A.Yu	1913-	mech eng		1960
Istrin, V.M.	1865-1937	lit historian	1902	1907
Istrina, Ye.S.	1883-1957	linguist	1943	
Ivanov, A.A.	1867-1939	astronomer	1925	
Ivanov, A.A.	1902-1956	geologist	1953	
Ivanov, L.A.	1871-1962	botanist	1922	
Ivanov, L.N.	1903-1957	historian	1939	1943
Ivanov, V.A.*	?	?	?	
Ivanov, V.Ye.	1908-	chemist	1964	
Iyerusalimskiy, N.D.	1901-1967	microbiologist	1960	1966
Iyevlev, V.M.	1923-	power eng	1964	
Izgaryshev, N.A.	1884-1956	electrochemist	1939	
Kabachnik, M.I.	1908-	organic chemist	1953	1958
Kabanov, V.A.	1934-	chemist	1968	
Kablukov, I.A.	1857-1942	physicochemist	1928	1932
Kachalov, N.N.	1883-1961	technol	1933	

* 1936 recorded as corresp member, USSR Acad of Sci.

Kadomtsev, B.B.	1928-	physicist	1962		
Kafarov, V.V.	1914-	chemical technol	1966		
Kalesnik, S.V.	1901-	geographer	1953	1968	
Kamenskiy, G.N.	1892-1959	hydro-geologist	1953		
Kammari, M.D.	1898-1965	philosopher	1953		
Kantorovich, L.V.	1912-	mathematician	1958	1964	
Kapelyushnikov, M.A.	1886-?	petroleum eng	1939		
Kapitsa, P.L.	1894-	physicist	1929	1939	
Kapustinskiy, A.F.	1906-1960	chemist	1939		
Karakeyev, K.-G.	1913-	historian	1968		
Karandeyev, K.B.	1907-1969	electr eng	1958		
Karavayev, N.M.	1890-	chemist	1946		
Kareyev, N.I.	1850-1931	historian	1910		1929
Kargapolov, M.I.	1928-	mathematician	1966		
Kargin, V.A.	1907-1969	physico-chemist	1946	1953	
Karnaukhov, M.M.	1892-1955	metallurgist	1939	1953	
Karpinskiy, A.P.	1847-1936	geologist	1889	1896	
Karskiy, Ye.F.	1861-1931	linguist	1901	1916	
Kazanskiy, B.A.	1891-	chemist	1943	1946	

Name	Born	Field		
Kazarnovskiy, I.A.	1890-	chemist	1939	
Kedrov, B.M.	1903-	philosopher	1960	1966
Keldysh, L.V.	1931-	physicist	1968	
Keldysh, M.V.	1911-	mathematician	1943	1946
Kell', N.G.	1883-1965	geodesist	1946	
Keller, B.A.	1874-1945	botanist, ecologist		1931
Kerimov, D.A.	1923-	lawyer	1966	
Khachapuridze (Pirozhkov),G.V.	1892-1957	historian	1939	
Khachaturov, T.S.	1906-	economist	1943	1966
Khain, V.Ye.	1914-	geologist, geophysicist	1966	
Khariton, Yu.B.	1904-	physico-chemist	1943	1953
Kharkevich, A.A.	1904-1965	radio eng	1960	1964
Khel'kvist, G.A.	1894-1968	geologist	1958	
Khimich, G.L.	1908-	mech eng	1968	
Khinchin, A.Ya.	1894-1959	mathematician	1939	
Khitarov, N.I.	1903-	geochemist	1964	
Khitrin, L.N.	1907-1965	thermal physicist	1953	
Khlopin, V.G.	1890-1950	chemist	1933	1939

Name	Dates	Field		
Khokhlov, A.S.	1916-	chemist	1964	
Khokhlov, R.V.	1926-	radio physicist	1966	
Khomentovskiy, A.S.	1908-	geologist	1960	
Khrapchenko, M.B.	1904-	lit historian	1958	1966
Khrenov, K.K.	1894-	electr welding specialist	1953	
Khristianovich, S.A.	1908-	mech eng	1939	1943
Khrushchov, G.K.	1897-1962	histologist	1953	
Khvol'son, O.D.	1852-1934	physicist	1895	1920
Khvostov, V.M.	1905-	historian	1953	1964
Kibel', I.A.	1904-	mathematician	1943	
Kikoin, I.K.	1908-	physicist	1943	1953
Kim, M.P.	1908-	historian	1960	
Kirenskiy, L.V.	1909-1969	physicist	1964	
Kirillin, V.A.	1913-	thermal eng	1953	1962
Kirpichev, M.V.	1879-1955	thermal eng	1929	1939
Kiselyov, S.V.	1905-1962	archeologist	1953	
Kishkin, S.T.	1906-	metallurgist	1960	1966
Kistyakovskiy, V.A.	1865-1952	physico-chemist	1925	1929

Kisun'ko, G.V.	1918-	radio eng	1958	
Kizhner, N.M.	1867-1935	chemist	1929	1934
Klimov, V.Ya.	1892-1962	mech eng	1943	1953
Knipovich, N.M.	1862-1939	zoologist	1927	1935
Knorre, D.G.	1926-	physico-chemist, biochemist	1968	
Knunyants, I.L.	1906-	organic chemist	1946	1953
Kobeko, P.P.	? -1954	physico-chemist	?	
Kobzarev, Yu.B.	1905	radio eng	1953	
Kocheshkov, K.A.	1894-	chemist	1946	1968
Kochetkov, N.K.	1915-	chemist	1960	
Kochin, N.Ye.	1901-1944	mathematician		1939
Kochina (Polubarinova-Kochina), P.Ya.	1899-	hydrodynamics specialist	1946	1958
Kokovtsev, P.K.	1861-1942	Semitist	1906	1912
Kolmogorov, A.N.	1903-	mathematician		1939
Kolosov, G.V.*	?	?	?	
Kolosov, M.N.	1927-	chemist	1966	
Kolosov, N.G.	1897-	histologist	1953	

---

* 1936 recorded as corresp member, USSR Acad of Sci.

Kolotyrkin, Ya.M.	1910-	physico-chemist	1966	
Kol'tsov, M.Ye.	1898-1942	writer	1938	
Kol'tsov, N.K.*	?	?	?	
Komarov, V.L.	1869-1945	botanist		1920
Kondakov, N.P.**	1844-1925	art historian	1892	1898
Kondrat'yev, K.Ya.	1920-	geophysicist	1968	
Kondrat'yev, V.N.	1902-	physico-chemist	1943	1953
Koni, A.F.	1844-1927	lawyer		1900
Konobeyev-skiy, S.T.	1890-1970	physicist	1946	
Kononov, A.N.	1906-	Turcologist	1958	
Konovalov, D.P.	1856-1929	chemist		1923
Konrad, N.I.	1891-1970	Orientalist	1934	1958
Konstantinov, B.P.	1910-1969	physicist	1953	1960
Konstantinov, F.V.	1901-	philosopher	1953	1964
Koptyug, V.A.	1931-	organic chemist	1968	
Korneychuk, A.Ye.	1905-	writer		1943

* 1935 recorded as corresp member, USSR Acad of Sci.
** Died in emigration in Prague.

Korolyov, S.P.	1907-1966	mech eng	1953	1958
Korotkov, A.A.	1910-1967	chemist	1958	
Korovin, Ye.A.	1892-1964	lawyer	1946	
Korshak, V.V.	1909-	chemist	1953	
Korzhinskiy, D.S.	1899-	geologist	1943	1953
Koshlyakov, N.S.	1891-1958	mathematician	1933	
Koshtoyants, Kh.S.	1900-1961	physiologist	1939	
Kosminskiy, Ye.A.	1886-1959	historian	1939	1946
Kostenko, M.P.	1889-	electr eng	1939	1953
Kostenko, M.V.	1912-	power eng	1962	
Kostinskiy, S.K.	1867-1936	astronomer	1915	
Kostyakov, A.N.	1887-1957	meliorator	1933	
Kostychyov, S.P.	1877-1931	plant physiologist	1922	1923
Kostyuk, P.G.	1924-	physiologist	1966	
Kosygin, Yu.A.	1911-	geologist	1958	
Kotel'nikov, V.A.	1908-	radio eng		1953
Kotlyarevskiy, N.A.	1863-1925	philologist		1909
Koton, M.M.	1908-	chemist	1960	
Kovalenkov, V.I.	1884-1960	wire communication specialist	1939	

Kovalyov, N.N.	1908-	mech eng	1953	
Koval'skiy, A.A.	1906-	physico-chemist	1958	
Kovda, V.A.	1904-	soil scientist	1953	
Kozin, S.A.	1879-1956	Mongolic specialist		1943
Kozlov, G.A.	1901-	economist	1968	
Kozlov, V.Ya.	1914-	mathematician	1966	
Kozo-Polyanskiy, B.M.	1890-1957	botanist	1932	
Kozyrev, B.M.	1905-	physicist	1968	
Krachkovskiy, I.Yu.	1883-1951	Orientalist		1921
Krasil'nikov, N.A.	1896-	microbiologist	1946	
Krasnovskiy, A.A.	1913-	biochemist	1962	
Krasovskiy, A.A.	1921-	control systems specialist	1968	
Krasovskiy, F.N.	1878-1948	geodesist	1939	
Krasovskiy, N.N.	1924-	mech eng	1964	1968
Krasuskiy, K.A.	1867-1937	chemist	1933	
Krats, K.O.	1914-	geologist	1968	
Kravets, T.P.	1876-1955	physicist	1943	
Kravkov, N.P.	1865-1924	pharmacologist	1920	

Kravkov, S.V.	1893-1951	philosopher	1946	
Kreps, Ye.M.	1899-	physiologist	1946	1966
Kretovich, V.L.	1907-	biochemist	1962	
Krishtofo-vich, A.N.	1885-1953	paleo-botanist	1953	
Kropotkin, P.N.	1910-	geologist	1966	
Krug, K.A.	1873-1952	electr eng	1933	
Krupskaya, N.K.	1869-1939	govt official		1930
Krutkov, Yu.A.	1890-1952	physicist	1933	
Kruus, Kh.Kh.	1891-	historian	1946	
Kruzhilin, G.N.	1911-	thermal eng	1953	
Kruzhkov, V.S.	1905-	philosopher	1953	
Krylov, A.N.	1863-1945	mech eng, shipbuilder	1914	1916
Krylov, A.P.	1904-	chemist, geologist	1953	1968
Krylov, N.M.	1879-1955	mathematician	1928	1929
Krylov, P.N.	1850-1931	botanist	1929	
Krzhizhanov-skiy, G.M.	1872-1959	power eng		1929
Kukin, D.M.	1908-	historian	1964	
Kulagin, N.M.	1860-1940	entomologist	1913	
Kulebakin, V.S.	1891-1970	electr eng	1933	1939

Kuleshov, P.N.	1854-1934	zootechnician	1928	
Kunin, V.N.	1906-	hydrogeologist	1968	
Kuprevich, V.F.	1897-1969	botanist	1953	
Kurchatov, I.V.	1903-1960	physicist		1943
Kurdyumov, G.V.	1902-	metals specialist	1946	1953
Kurnakov, N.S.	1860-1941	chemist		1913
Kursanov, A.L.	1902-	physiologist, biochemist	1946	1953
Kursanov, D.N.	1899-	chemist	1953	
Kutateladze, S.S.	1914-	power eng	1968	
Kuusinen, O.V.	1881-1964	historian		1958
Kuzin, A.M.	1906-	biophysicist	1960	
Kuz'min, R.O.	1891-1949	mathematician	1946	
Kuznetsov, N.D.	1911-	mech eng	1968	
Kuznetsov, N.I.	1864-1932	botanist	1904	
Kuznetsov, S.I.	1900-	microbiologist	1960	
Kuznetsov, V.A.	1906-	geologist	1958	
Kuznetsov, V.D.	1887-1963	physicist	1946	1958
Kuznetsov, V.I.	1913-	mech eng	1958	1968

Kuznetsov, Yu.A.	1903-	geologist	1958	1966
Lagorio, A.Ye.	1852-?	petrographer	1896	
Landau, L.D.	1908-1968	physicist		1946
Landsberg, G.S.	1890-1957	physicist	1932	1946
Lapo-Danilev-skiy, A.S.	1863-1919	historian	1902	1905
Lapo-Danilev-skiy, I.A.	1895-1931	mathematician	1931	
Larionov, A.N.	1889-1963	electr eng	1953	
Laskorin, B.N.	1915-	chemical technol	1966	
Latyshev, V.V.	1855-1921	philologist, historian	1890	1893
Lavochkin, S.A.	1900-1960	plane designer	1958	
Lavrenko, Ye.M.	1900-	geobotanist	1946	1968
Lavrent'yev, B.I.	1892-1944	histologist	1939	
Lavrent'yev, M.A.	1900-	mathematician		1946
Lavrent'yev, M.M.	1932-	mathematician	1968	
Lavrov, S.S.	1923-	automatic control systems specialist	1966	
Lavrovskiy, K.P.	1898-	chemist	1953	
Lazarev, P.P.	1878-1942	physicist, biophysicist		1917

Name	Born–Died	Profession		
Lazarev, V.N.	1897-	art historian	1943	
Lebedev, A.A.	1893-1969	physicist	1939	1943
Lebedev, A.B.	1883-1941	electr traction eng	1939	
Lebedev, P.I.	1885-1948	geologist, petrographer	1939	
Lebedev, S.A.	1902-	electr eng		1953
Lebedev, S.V.	1874-1934	chemist	1939	
Lebedev-Polyanskiy, P.I.	1881-1948	lit historian	1939	1946
Lebedinskiy, V.V.	1888-1956	chemist	1946	
Lein'sh, P.Ya.	1883-1959	biologist	1946	
Leont'yev, L.A.	1901-	economist	1939	
Leontovich, M.A.	1903-	physicist	1939	1946
Letov, A.M.	1911-	control theory specialist	1968	
Levich, V.G.	1917-	physicochemist	1958	
Levina, R.S.	1899-1964	economist	1939	
Levinson-Lessing, F.Yu.	1861-1939	geologist	1914	1925
Levitskiy, O.D.	1909-1961	geologist	1953	
Levitskiy, R.A.	?	biologist	1932	
Levkoyev, I.I.	1909-	chemist	1968	

Leybenzon, L.S.	1879-1951	geophysicist	1933	1943
Lidorenko, N.S.	1916-	electr eng	1966	
Lifshits, I.M.	1917-	physicist	1960	
Lifshits, Ye.M.	1915-	physicist	1966	
Likhachyov, D.S.	1906-	lit historian	1953	
Likhachyov, N.P.	1862-1935	historian		1925
Linnik, V.P.	1889-	physicist		1939
Linnik, Yu.V.	1915-	mathematician	1953	1964
Lipin, V.N.	1858-1930	metallurgist	1928	
Lipskiy, V.I.	1863-1937	botanist	1928	
Livanov, M.N.	1907-	physiologist	1962	
Logunov, A.A.	1926-	physicist	1968	
Luchitskiy, I.V.	1912-	geologist	1968	
Lukin, N.M.	1885-1940	historian		1929
Lukirskiy, P.I.	1894-1954	physicist	1933	1946
Luk'yanenko, P.P.	1901-	geneticist		1964
Lunacharskiy, A.V.	1875-1933	lit critic		1930
Luppol, I.K.	1896-1943	philosopher	1933	1939
Lur'ye, A.I.	1901-	mech eng	1960	
Luzin, N.N.	1883-1950	mathematician	1927	1929

L'vov, S.D.	1879-1959	physiologist, biochemist	1946
Lyapunov, A.A.	1911-	mathematician	1964
Lyapunov, A.M.	1857-1918	mathematician	1900 1901
Lyapunov, B.M.	1862-1943	Slavist	1923
Lyashchenko, P.I.	1876-1955	economist	1943
Lysenko, T.D.	1898-	agronomist	1939
Lyubavskiy, M.K.	1860-1936	historian	1917 1929
Lyubimenko, V.N.	1873-1937	botanist, physiologist	1922
Lyul'ka, A.M.	1908-	plane eng specialist	1960 1968
Lyusternik, L.A.	1899-	mathematician	1946
Magnitskiy, V.A.	1915-	geophysicist	1964
Makarevskiy, A.I.	1904-	plane designer	1953 1968
Makeyev, V.P.	1924-	mech eng	1968
Makovel'skiy, A.O.	1884-1969	philosopher	1946
Maksimov, A.A.	1891-	philosopher	1943
Maksimov, N.A.	1880-1952	plant physiologist	1932 1946
Maksutov, D.D.	1896-1964	astrooptics specialist	1946
Malein, A.K.	1869-1938	historian	1916

Malov, S.Ye.	1880-1957	Turcologist	1939	
Mal'tsev, A.I.	1909-1967	mathematician	1953	1958
Malyusov, V.A.	1913-	technol chemist	1968	
Mamedaliyev, Yu.G.	1905-1961	chemist	1958	
Manandyan, Ya.A.	1873-1952	historian		1939
Mandel'shtam, L.I.	1879-1944	physicist	1928	1929
Man'kovskiy, G.I.	1897-1965	mining eng	1960	
Marchuk, G.I.	1925-	mathematician	1962	1968
Mardzhanash-vili, K.K.	1903-	mathematician	1964	
Markov, A.A.	1856-1922	mathematician	1890	1896
Markov, A.A.	1903-	mathematician	1953	
Markov, D.F.	1913-	lit historian	1966	
Markov, M.A.	1908-	physicist	1953	1966
Marr, N.Ya.	1864-1934	philologist		1912
Maslov, P.P.	1867-1946	economist		1929
Matulis, Yu.Yu.	1899-	chemist	1946	
Mayskiy, I.M.	1884-	historian, diplomat		1946
Medvedev, S.S.	1891-1970	chemist	1943	

Melent'yev, L.A.	1908-	power eng.	1960	1966
Melikishvili (Melikov), P.G.	1850-1927	chemist	1927	
Meller, G.Dzh.*	?	geneticist	1933	
Mel'nikov, N.V.	1909-	mining eng	1953	1962
Mel'nikov, O.A.	1912-	astronomer	1960	
Mel'nikov, P.I.	1908-	geologist	1968	
Menner, V.V.	1905-	geologist, geophysicist		1966
Men'shov, D.Ye.	1892-	mathematician	1953	
Menzbir, M.A.	1855-1935	zoologist	1896	1929
Mergelyan, S.N.	1928-	mathematician	1953	
Meshchaninov, I.I.	1883-1967	linguist, archeologist		1932
Meshcheryakov, M.G.	1910-	physicist	1953	
Meshcheryakov, N.L.	1865-1942	professional revolutionary	1939	
Meysel', M.N.	1901-	microbiologist	1960	
Michurin, I.V.	1855-1935	biologist		1935
Migdal, A.B.	1911-	physicist	1953	1966
Mikhalevskiy, F.I.	1876-1952	economist	1946	
Mikhaylov, A.A.	1888-	astronomer	1943	1964

* 1949 deprived of corresp membership under Paragraph 24 of the USSR Academy of Sciences' Statutes for "activities directed against the USSR."

Mikhaylov, B.M.	1906-	chemist	1968	
Mikheyev, M.A.	1902-1970	thermal eng	1946	1953
Mikoyan, A.I.	1905-1970	plane designer	1953	1968
Mikulin, A.A.	1895-	plane designer		1943
Mikulinskiy, S.R.	1919-	philosopher	1968	
Mileykovskiy, A.G.	1911-	economist	1966	
Miller, ??	?	biologist	1933	
Millionshchikov, M.D.	1913-	physicist	1953	1962
Mints, A.L.	1895-	radio eng	1946	1958
Mints, I.I.	1896-	historian	1939	1946
Mirchink, M.F.	1901-	geologist	1953	
Mironov, S.I.	1883-1959	geologist		1946
Mishin, V.P.	1917-	mech eng	1958	1966
Mishustin, Ye.N.	1901-	microbiologist	1953	
Mitin, M.B.	1901-	philosopher		1939
Mitkevich, V.F.	1872-1951	electr eng	1927	1929
Modzalevskiy, B.L.	1874-1928	lit historian	1918	
Moiseyev, N.N.	1917-	mech eng	1966	
Molchanov, A.A.	1902-	timber ind specialist	1968	

Molodenskiy, M.S.	1909-	geophysicist	1946	
Molotov (Skryabin), V.M.*	1890-	govt official		1946
Morozov, N.A.	1854-1946	professional revolutionary		1932
Moshkin, P.A.	1891-	technol chemist	1953	
Muratov, M.V.	1908-	geologist	1962	
Muskhelishvili, N.I.	1891-	mathematician	1933	1939
Mustel', E.R.	1911-	astrophysicist	1953	
Nadson, G.A.**	?	microbiologist	?	
Nalivkin, D.V.	1889-	geologist	1933	1946
Nalivkin, V.D.	1915-	geologist	1968	
Namyotkin, N.S.	1916-	chemist	1962	
Namyotkin, S.S.	1876-1950	organic chemist	1932	1939
Nasonov, D.N.	1895-1957	cytophysiologist	1943	
Nasonov, N.V.	1855-1939	zoologist	1897	1906
Naumov, A.A.	1916-	physicist	1964	
Naumov, N.A.	1888-1959	botanist	1946	

---

* On 26 Mar 1959 the Academy's Gen Assembly deprived him of hon membership for belonging to a "schismatic group."

** 1932 recorded as corresp member, USSR Acad of Sci.

Navashin, S.G.	1857-1930	biologist	1901	1918
Nazarov, I.N.	1906-1957	organic chemist	1946	1953
Nechkina, M.V.	1901-	historian	1953	1958
Nekrasov, A.I.	1883-1957	mech eng	1932	1946
Nekrasov, B.V.	1899-	chemist	1946	
Nekrasov, N.N.	1906-	economist	1958	1968
Nemchinov, V.S.	1894-1964	economist		1946
Nenadkevich, K.A.	1880-1963	chemist, mineralogist	1946	
Nesmeyanov, A.N.	1899-	organic chemist	1939	1943
Neyman, L.R.	1902-	electr eng	1953	
Nikiforov, P.M.	1884-1944	geophysicist	1932	
Nikitin, N.I.	1890-	chemist	1939	
Nikitin, V.P.	1893-1956	electr eng		1939
Nikitskiy, A.V.	1859-1921	historian	1902	1917
Nikolayev, A.V.	1902-	chemist	1958	1966
Nikolayev, I.I.	1893-1964	transport eng	1953	
Nikolayev, V.A.	1893-1960	petrographer	1946	
Nikol'skiy, B.P.	1900-	physico-chemist	1953	1968

Nikol'skiy, G.V.	1910-	ichthyologist	1953	
Nikol'skiy, N.K.	1863-1936	lit historian	1900	1916
Nikol'skiy, N.M.	1877-1959	Orientalist	1946	
Nikol'skiy, S.M.	1905-	mathematician	1968	
Novikov, I.I.	1916-	physicist	1958	
Novikov, P.S.	1901-	mathematician	1953	1960
Novikov, S.P.	1938-	mathematician	1966	
Novosyolova, A.V.	1900-	chemist	1953	
Novozhilov, V.V.	1910-	mech eng	1958	1966
Numerov, B.V.*	1891-1941	astronomer	?	
Nuzhdin, N.I.	1904-	biologist	1953	
Obnorskiy, S.P.	1888-1962	linguist		1939
Obraztsov, I.F.	1920-	mech eng	1966	
Obraztsov, V.N.	1874-1949	railroad eng		1939
Obreimov, I.V.	1894-	physicist	1933	1958
Obruchev, S.V.	1891-1965	geologist	1953	
Obruchev, V.A.	1863-1956	geologist, geographer	1921	1929
Obukhov, A.M.	1918-	geophysicist	1953	
Oding, I.A.	1896-1964	metallurgist	1946	

* 1936 recorded as corresp member, USSR Acad of Sci.

Odintsov, M.M.	1913-	geologist	1964	
Okhotsimskiy, D.Ye.	1921-	mech eng	1960	
Okladnikov, A.P.	1908-	historian, archeologist	1964	1968
Oksman, Yu.G.*	?	?	?	
Okun', L.B.	1929-	physicist	1966	
Ol'denburg, S.F.	1863-1934	Orientalist	1903	1908
Ol'derogge, D.A.	1903-	linguist	1960	
Omel'yanovskiy, M.E.	1904-	philosopher	1968	
Omelyanskiy, V.L.	1867-1928	microbiologist	1916	1923
Oparin, A.I.	1894-	biochemist	1939	1946
Orbeli, I.A.	1887-1961	Orientalist	1932	1935
Orbeli, L.A.	1882-1958	physiologist	1932	1935
Orekhov, A.P.	1881-1939	chemist		1939
Oreshnikov, A.V.	1855-1933	historian	1928	
Orlov, A.S.	1871-1947	lit historian		1931
Orlov, A.Ya.	1880-1954	astronomer	1927	
Orlov, S.V.	1880-1958	astronomer	1943	

---

* 1937 deprived of corresp membership as an "enemy of the people."

Orlov, Yu.A.	1893-1966	paleontologist	1953	1960
Orlovskiy, P.Ye.	1896-	lawyer	1953	
Osinskiy (Obolenskiy), V.V.*	1887-1938	economist	?	
Osipov, V.P.	1871-1947	psychiatrist	1939	
Ossovskiy, A.V.	1871-1957	musicologist	1943	
Ostrovityanov, K.V.	1892-1969	economist	1939	1953
Ovchinnikov, L.N.	1913-	geologist	1964	
Ovchinnikov, Yu.A.	1934-	chemist	1968	
Ovsyannikov, L.V.	1919-	mech eng	1964	
Oyzerman, T.I.	1914-	philosopher	1966	
Palladin, A.V.	1885-	biochemist		1942
Palladin, V.I.	1859-1922	plant physiologist	1905	1914
Pal'mov, I.S.	1856-1920	philologist	1913	1916
Pankratova, A.M.	1897-1957	historian		1953
Papaleksi, N.D.	1880-1947	physicist	1931	1939
Papkovich, P.F.	1887-1946	shipbuilder	1933	

---

* 1932 recorded as full member, USSR Acad of Sci. Arrested during personality cult. Posthumously rehabilitated.

Parenago, P.P.	1906-1960	astronomer	1953		
Parin, V.V.	1903-	physiologist		1966	
Pariyskiy, N.N.	1900-	geophysicist	1968		
Parland, A.A.	1842-1920	architect		1881	
Parnas, Ya.O.	1884-1949	biochemist		1942	
Pashkov, A.I.	1900-	economist	1953		
Paton, B.Ye.	1918-	electr welding specialist		1962	
Pavlov, A.P.	1854-1929	geologist	1905	1916	
Pavlov, I.M.	1900-	metallurgist	1946		
Pavlov, I.P.	1849-1936	physiologist	1901	1907	
Pavlov, M.A.	1863-1958	metallurgist	1927	1932	
Pavlova, M.V.	1854-1938	paleontologist	1925		1930
Pavlovskiy, N.N.	1884-1937	hydraulic eng		1932	
Pavlovskiy, Ye.N.	1884-1965	zoologist, parasitologist		1939	
Pazhitnov, K.A.	1879-1964	economist, historian	1946		
Pekarskiy, E.K.	1858-1934	ethnographer	1927		1931
Perederiy, G.P.	1871-1953	civil eng	1939	1943	

Perets, V.N.	1870-1936	lit historian		1914	
Petrenko-Kritchenko, P.I.	1866-1944	organic chemist	1932		
Petrov, A.A.	1913-	organic chemist	1966		
Petrov, A.D.	1895-1964	chemist	1946		
Petrov, A.P.	1910-	rail transport eng	1953		
Petrov, B.N.	1913-	automation and telemech specialist	1953	1960	
Petrov, G.I.	1912-	hydroaeromechanic	1953	1958	
Petrov, N.N.	1876-1964	surgeon, oncologist	1939		
Petrov, N.P.	1836-1920	hydrodynamicist			1894
Petrovskiy, B.V.	1908-	surgeon		1966	
Petrovskiy, I.G.	1901-	mathematician	1943	1946	
Petrushevskiy, D.M.	1863-1942	historian	1924	1929	
Petryanov-Sokolov, I.V.	1907-	physicochemist	1953	1966	
Peyve, A.V.	1909-	geologist	1958	1964	
Peyve, Ya.V.	1906-	agrochemist, plantbreeder	1953	1966	
Picheta, V.I.	1878-1947	historian	1939	1946	
Pigulevskaya, N.V.	1894-1970	historian	1946		

Piksanov, N.K.	1878-1969	lit historian	1931	
Pilyugin, N.A.	1908-	mech eng	1958	1966
Piontkovskiy, A.A.	1898-	lawyer	1968	
Pisarzhevskiy, L.V.	1874-1938	chemist	1928	1930
Pistol'kors, A.A.	1896-	radio eng	1946	
Piyp, B.I.	1906-1966	vulcanologist	1958	
Plaksin, I.N.	1900-1967	mining eng	1946	
Platonov, S.F.*	1860-1933	historian	1908	1920
Plaude, K.K.	1897-	thermal eng	1960	
Plotnikov, K.N.	1907-	economist	1960	
Plotnikov, V.A.	1873-1947	organic chemist	1932	
Podvysotskaya, O.N.	1884-1958	dermatologist	1939	
Pogorelov, A.V.	1919-	mathematician	1960	
Pokrovskiy, M.M.	1869-1942	linguist		1929
Pokrovskiy, M.N.	1868-1932	historian		1929
Polkanov, A.A.	1888-1963	geologist		1943

---

* Held monarchist views. 1931 deprived of membership of Academy. Arrested and exiled to Samara. Died in exile.

Polyakov, Yu.A.	1921-	historian	1966	
Polynov, B.B.	1877-1952	soil scientist, geochemist	1933	1946
Pomeranchuk, I.Ya.	1913-1966	physicist	1953	1964
Ponomaryov, B.N.	1905-	historian	1958	1962
Pontekorvo, B.M.	1913-	physician	1958	1964
Pontryagin, L.S.	1908-	mathematician	1939	1958
Popkov, V.I.	1908-	electr eng	1953	1966
Popkovich, P.F.*	?	?	?	
Popov, V.V.	1902-1960	entomologist	1953	
Popov, Ye.P.	1914-	mech eng	1960	
Poppe, N.N.	?	Orientalist	1931	
Poray-Koshits, A.Ye.	1877-1949	chemist	1931	1935
Pospelov, G.S.	1914-	automatic control systems specialist	1966	
Pospelov, P.N.	1898-	historian	1946	1953
Posse, K.A.	1847-1928	mathematician		1916
Potyomkin, F.V.	1895-	historian	1953	
Potyomkin, V.P.	1878-1946	historian		1943
Pozdyunin, V.L.	1883-1948	shipbuilder		1939

* 1932 recorded as corresp member, USSR Acad of Sci.

Prasolov, L.I.	1875-1954	geographer, soil scientist	1931	1935
Predvoditelev, A.S.	1891-	physicist	1939	
Priklonskiy, V.A.	1899-1959	hydro-geologist	1958	
Prilezhayev, N.A.	1872-1944	organic chemist	1933	
Privalov, I.I.	1891-1941	mathematician	1939	
Prokhorov, A.M.	1916-	physicist	1960	1966
Prokhorov, Yu.V.	1929-	mathematician	1966	
Prokof'yev, M.A.	1910-	organic chemist	1966	
Prudenskiy, G.A.	1904-1967	economist	1958	
Pryanishnikov, D.N.	1865-1948	agrochemist	1913	1929
Pshenitsyn, N.K.	1891-1961	chemist	1953	
Ptitsyn, B.V.	1903-1965	chemist	1960	
Ptukha, M.V.	1884-1961	statistician	1943	
Pudovik, A.N.	1916-	organic chemist	1964	
Pugachyov, V.S.	1911-	cyberneticist	1966	
Pustovalov, L.V.	1902-	petrographer	1953	
Pustovoyt, V.S.	1886-	geneticist		1964
Puzyryov, N.N.	1914-	geophysicist	1966	

Rabinovich, A.I.	1893-1942	physico-chemist	1933	
Rabinovich, I.M.	1886-	mech eng	1946	
Rabotnov, Yu.N.	1914-	mech eng	1953	1958
Radlov, V.V.	1837-1918	Turcologist	1884	
Radzig, A.A.	1869-1941	thermal power eng	1933	
Rakitin, Yu.V.	1911-	plant physiologist	1962	
Rakovskiy, A.V.	1879-1941	physico-chemist	1933	
Raspletin, A.A.	1908-1967	radio eng	1958	1964
Raushenbakh, B.V.	1915-	mech eng, cyberneticist	1966	
Ravdonikas, V.I.	1894-	historian, archeologist	1946	
Razin, N.V.	1904-	hydraulic power eng	1968	
Razuvayev, G.A.	1895-	organic chemist	1958	1966
Rebinder, P.A.	1898-	physico-chemist	1933	1946
Reformatskiy, S.N.	1860-1934	chemist	1928	
Regel', V.E.	1857-1932	historian	1898	
Rengarten, V.P.	1882-1964	geologist	1946	
Reutov, O.A.	1920-	chemist	1958	1964

Rikhter, A.A.	1871-1947	botanist, physiologist	1929	1932
Riznichenko, Yu.V.	1911-	geophysicist	1958	
Rodionov, V.M.	1878-1954	chemist	1939	1943
Roginskiy, S.Z.	1900-1970	physico-chemist	1939	
Romankov, P.G.	1904-	technol chemist	1964	
Romashkin, P.S.	1915-	lawyer	1958	
Ronov, A.B.	1913-	geochemist	1966	
Rotshteyn, F.A.	1871-1953	historian	1939	
Roytbak, A.I.	1919-	physiologist	1968	
Rozanov, M.N.	1858-1936	lit historian		1921
Rozenberg, D.I.	1879-1950	economist	1939	
Rozhanskiy, D.A.	1882-1936	physicist	1933	
Rozhdestven-skiy, D.S.	1876-1940	physicist	1925	1929
Rozhdestven-skiy, S.V.	1868-1934	historian	1920	
Rozhkov, I.S.	1908-	geologist	1960	
Rubinshteyn, S.L.	1889-1960	philosopher, psychologist	1943	
Rumyantsev, A.M.	1905-	economist	1960	1966
Ryabchikov, D.I.	1904-1965	analytical chemist	1964	

Ryabushkin, T.V.	1914-	economist	1966	
Ryazanov, D.B.	1870-1938	historian		1929
Ryazanskiy, M.S.	1909-	radio eng	1958	
Rybakov, B.A.	1908-	historian, archeologist	1953	1958
Rykachev, M.A.	1841-1919	meteorologist	1892	1900
Rykalin, N.N.	1903-	welding specialist	1953	1968
Ryl'skiy, M.F.	1895-1964	poet		1958
Rytov, S.M.	1908-	physicist	1968	
Ryzhikov, K.M.	1912-	biologist	1964	
Ryzhkov, V.L.	1896-	biologist	1946	
Rzhanov, A.V.	1920-	physicist	1962	
Rzhevskiy, V.V.	1913-	mining eng	1966	
Sadovskiy, M.A.	1904-	physicist	1953	1966
Sadovskiy, V.D.	1908-	technologist	1968	
Sadykov, A.S.	1913-	organic chemist	1966	
Sagdeyev, R.Z.	1932-	physicist	1964	1968
Sakharov, A.D.	1921-	physicist		1953
Saks, V.N.	1911-	geologist, geomorphologist	1958	
Sakulin, P.N.	1868-1930	lit historian		1929
Samarin, A.M.	1902-1970	metallurgist	1946	1966

Samarskiy, A.A.	1919-	mathematician	1966	
Samsonov, A.M.	1908-	historian	1964	
Samoylovich, A.N.	?	Orientalist		1929
Sapozhnikov, L.M.	1906-1970	thermal eng	1946	
Satkevich, A.A.	?	?	1933	
Satpayev, K.I.	1899-1964	geologist		1946
Saukov, A.A.	1902-1964	geochemist	1953	
Savarenskiy, F.P.	1881-1946	hydro-geologist	1939	1943
Savarenskiy, Ye.F.	1911-	geologist, geophysicist	1966	
Savel'yev, M.A.	1884-1939	Party official		1932
Savitskiy, Ye.M.	1912-	metallurgist	1966	
Sazhin, N.P.	1897-1969	metallurgist	1953	1964
Sedov, L.I.	1907-	mech eng	1946	1953
Semenikhin, V.S.	1918-	automation and telemech specialist	1968	
Semyonov, N.N.	1896-	physicist	1929	1932
Serebrennikov, B.A.	1915-	linguist	1953	
Serebrovskiy, B.S.	?	biologist	1933	
Sergeyev, Ye.M.	1914-	hydro-geologist	1966	

Sergeyev-Tsenskiy, S.N.	1875-1958	writer		1943
Severin, S.Ye.	1901-	biochemist	1953	1968
Severnyy, A.B.	1913-	astronomer	1958	1968
Severtsov, A.N.	1866-1936	biologist		1920
Shafarevich, I.R.	1923-	mathematician	1958	
Shakhmatov, A.A.	1864-1920	linguist	1897	1899
Shakhov, F.N.	1894-	geologist	1958	
Shaksel', Yu.Yu.*	?	?	?	
Shal'nikov, A.I.	1905-	physicist	1946	
Shanidze, A.G.	1887-	linguist	1939	
Shapiro, F.L.	1915-	physicist	1968	
Shaposhnikov, V.N.	1884-1968	microbiologist		1953
Shatelen, M.A.	1866-1957	electr eng	1931	
Shatskiy, N.S.	1895-1960	geologist	1943	1953
Shayn, G.A.	1892-1956	astronomer		1939
Shchedro, N.K.**	?	?	?	
Shcheglyayev, A.V.	1902-	thermal eng	1953	

---

* 1935 recorded as corresp member, USSR Acad of Sci.
** 1936 recorded as corresp member, USSR Acad of Sci.

Shchyolkin, K.I.	1911-1968	physicist	1953	
Shcherba, L.V.	1880-1944	linguist	1924	1943
Shcherbakov, D.I.	1893-1966	geologist, geochemist	1946	1953
Shcherbatskoy, F.I.	1866-1942	Indologist		1918
Shchukin, A.N.	1900-	radio eng	1946	1953
Shchusev, A.V.	1873-1949	architect		1943
Shemyakin, M.M.	1908-1970	organic chemist	1953	1958
Shenfer, K.I.	1885-1946	electr eng	1931	1932
Shennikov, A.P.	1888-1962	botanist	1946	
Shepilov, D.T.*	1905-	economist	1953	
Shestakov, A.V.	1877-1941	historian	?	
Shestakov, S.P.	1864-1940	historian, philologist	1916	
Shevyakov, L.D.	1889-1963	mining eng		1939
Sheyndlin, A.Ye.	1916-	power eng	1964	
Shilo, N.A.	1913-	geologist	1964	
Shimanskiy, Yu.A.	1883-1962	shipbuilder	1933	1953

---

* On 26 Mar 1959 deprived of corresp membership by the Academy's Gen Assembly for alleged membership of a "schismatic group."

Shimkevich, V.M.	1858-1923	zoologist		1920	
Shirkov, D.V.	1928-	physicist	1960		
Shirshov, A.I.	1921-	mathematician	1964		
Shirshov, P.P.	1905-1953	hydro-biologist		1939	
Shishkin, B.K.	1886-1963	botanist	1943		
Shishmarev, V.F.	1874-1957	philologist	1924	1946	
Shklovskiy, I.S.	1916-	astro-physicist	1966		
Shlyk, A.A.	1928	physiologist	1966		
Shmal'gauzen, I.I.	1884-1963	zoologist		1935	
Shmidt, A.E.	1871-1939	Islamist	1925		
Shmidt, O.Yu.	1891-1956	mathematician, astronomer	1933	1935	
Shnirel'man, L.G.	1905-1938	mathematician	1933		
Shokal'skiy, Yu.M.	1856-1940	oceanographer	1925		1939
Sholokhov, M.A.	1905-	writer		1939	
Shorygin, P.P.	1881-1939	organic chemist	1932	1939	
Shostakovskiy, M.F.	1905-	organic chemist	1960		
Shpak, V.S.	1909-	technol chemist	1968		

Name	Dates	Field		
Shpil'reyn, Ya.N.*	?	?	?	
Shtelling, E.V.	1850 1922	geophysicist	1901	
Shtern, L.S.	1878-1968	physiologist		1939
Shternberg, L.Ya.	1861-1927	ethnographer	1924	
Shteynberg, S.S.	1872-1940	metallurgist	1939	
Shubnikov, A.V.	1887-1970	crystallographer	1933	1953
Shukhov, V.G.	1853-1939	eng	1927	1929
Shuleykin, M.V.	1884-1939	radio eng	1933	1939
Shuleykin, V.V.	1895-	geophysicist	1929	1946
Shunkov, V.I.	1900-1967	historian	1962	
Shuykin, N.I.	1898-1968	organic chemist	1953	
Shvarts, S.S.	1919-	zoologist	1966	
Shvetsov, P.F.	1910-	permafrost specialist	1953	
Sibirtsev, Yu.M.	1853-1933	historian, paleographer	1928	
Sidorenko, A.V.	1917-	geologist	1953	1966
Sidorov, A.A.	1891-	art historian	1946	
Sidorov, V.A.	1930-	physicist	1968	

---

* 1933 recorded as corresp member, USSR Acad of Sci.

251

Name	Born- Died	Field		
Siforov, V.I.	1904-	radio eng	1953	
Sil'ven Levi*	?	?	?	
Simoni, P.K.	1859-1939	lit historian	1939	
Sisakyan, N.M.	1907-1966	biochemist	1953	1960
Skazkin, S.D.	1890-	historian	1943	1958
Skobel'tsyn, D.V.	1892-	physicist	1939	1946
Skochinskiy, A.A.	1874-	mining eng		1935
Skrinskiy, A.N.	1936-	physicist	1968	
Skryabin, G.K.	1917-	biochemist	1968	
Skryabin, K.I.	1878-	helminthologist		1939
Slavyanov, N.N.	1878-1958	hydro-geologist	1946	
Slin'ko, M.G.	1914-	chemist	1966	
Smirnov, A.I.	1888-1945	plant physiologist	1943	
Smirnov, N.V.	1900-1966	mathematician	1960	
Smirnov, S.S.	1895-1947	geologist	1939	1943
Smirnov, V.I.	1910-	geologist	1958	1962
Smirnov, V.I.	1887-	mathematician	1932	1943
Smirnov, V.S.	1915-	metallurgist	1960	

* 1933 recorded as corresp member, USSR Acad of Sci.

Smirnov, Ya.I.	1869-1918	archeologist	1907	1917
Smit-Fal'kner, M.N.	1878-1968	economist	1939	
Sobolev, S.L.	1908-	mathematician	1933	1939
Sobolev, V.S.	1908-	petrographer		1958
Sobolev, V.V.	1915-	astronomer	1958	
Sobolevskiy, A.I.	1856-1929	philologist	1893	1900
Sobolevskiy, S.I.	1864-1963	philologist	1928	
Sochava, V.B.	1905-	geobotanist	1958	1968
Sokolov, A.V.	1898-	agrochemist	1964	
Sokolov, B.S.	1914-	paleontologist	1958	1968
Sokolov, S.Ya.	1897-1957	physicist	1953	
Sokolovskiy, V.V.	1912-	mech eng	1946	
Solntsev, S.I.	1872-1936	economist		1929
Solodovnikov, V.G.	1918-	economist	1966	
Solonenko, V.P.	1916-	geophysicist	1966	
Soloukhin, R.I.	1930-	mech eng	1968	
Sorokin, G.M.	1910-	economist	1962	
Sotskov, B.S.	1908-	automation specialist	1960	
Spasokukotskiy, S.I.	1870-1943	surgeon		1942

Speranskiy, A.D.	1888-1961	pathophysiologist		1939
Speranskiy, G.N.	1873-1969	pediatrician	1943	
Speranskiy, M.N.	1863-1938	lit historian		1921
Spirin, A.S.	1931-	biochemist	1966	
Spitsyn, V.I.	1902-	chemist	1946	1958
Spivak, P.Ye.	1911-	physicist	1964	
Spivakovskiy, A.O.	1888-	transport specialist	1946	
Sretenskiy, L.N.	1902-	mech eng, mathematician	1939	
Sreznevskiy, V.I.	1867-1936	lit historian	1907	
Stalin, I.V.	1879-1953	Party and govt official		1946
Stanislavskiy, K.S.	1863-1938	stage dir, actor		1937
Starik, I.Ye.	1902-1964	chemist	1946	
Stark, B.V.	1883-1955	metallurgist	1943	
Starovskiy, V.N.	1905-	economist	1958	
Stechkin, B.S.	1891-1969	thermal eng	1946	1953
Steklov, V.A.	1864-1926	mathematician	1902	1912
Stepanov, A.V.	1908-	physician	1968	
Stepanov, N.I.	1879-1938	chemist	1929	

Stepanov, P.I.	1880-1947	geologist		1939
Stepanov, V.V.	1889-1950	mathematician	1946	
Stepanov, V.Ye.	1913-	astro-physicist	1968	
Stepanyan, Ts.A.	1911-	philosopher	1964	
Strakhov, N.M.	1900-	geologist	1946	1953
Strazhesko, N.D.	1876-1952	therapist		1943
Streletskiy, N.S.	1885-1967	specialist in metal structures	1931	
Strelkov, P.G.	1899-1968	physicist	1960	
Strogovich, M.S.	1894-	lawyer	1939	
Strumilin, S.G.	1877-	economist		1931
Struminskiy, V.V.	1914-	mech eng	1958	1966
Struve, V.V.	1889-1965	historian		1935
Styrikovich, M.A.	1902-	thermal eng	1946	1964
Subbotin, M.F.	1893-1966	astronomer	1946	
Subbotin, V.I.	1919-	thermo-physicist	1968	
Suchkov, B.L.	1917-	lit historian	1968	
Sukachyov, V.N.	1880-1967	botanist, silviculturist	1920	1943
Sushkin, P.P.	1868-1928	zoologist		1923

Svetovidov, A.N.	1903-	ichthyologist	1953	
Svishchev, G.P.	1912-	mech eng	1966	
Syrkin, Ya.K.	1894-	physico-chemist	1943	1964
Syromyatnikov, S.P.	1891-1951	thermal eng		1943
Takhtadzhyan, A.L.	1910-	botanist	1966	
Talanov, V.V.	?	biologist	1932	
Talmud, D.L.	1900-	physico-chemist	1934	
Tal'roze, V.L.	1922-	chemophysicist	1968	
Tamm, I.Ye.	1895-	physicist	1933	1953
Tananayev, I.V.	1904-	chemist	1946	1958
Tanner, V.	?	geologist	1934	
Tarle, Ye.V.	1875-1955	historian	1921	1927
Tatarinov, P.M.	1895-	geologist	1953	
Tauson, L.V.	1917-	geochemist	1966	
Terenin, A.N.	1896-1967	physico-chemist	1932	1939
Terent'yev, A.P.	1891-1970	chemist	1953	
Terpigorev, A.M.	1873-1959	mining eng		1935
Terskov, I.A.	1918-	biophysicist	1968	
Tikhomirov, M.N.	1893-1965	historian	1946	1953

Tikhomirov, V.V.	1912-	radio eng	1953	
Tikhonov, A.N.	1906-	mathematician, geophysicist	1939	1966
Tikhov, G.A.	1875-1960	astronomer	1939	
Tikhvinskiy, S.L.	1918-	historian	1968	
Timakov, V.D.	1905-	epidemiologist		1968
Timiryazev, K.A.	1843-1920	botanist	1890	
Timofeyev, L.I.	1904-	lit historian	1958	
Timofeyev, P.V.	1902-	electronics specialist	1953	
Timofeyev, T.T.	1928-	economist	1966	
Tishchenko, V.Ye.	1861-1941	chemist	1928	1935
Tolstov, S.P.	1907-	archeologist, historian	1953	
Tolstoy, A.N.	1883-1945	writer		1939
Tolstoy, I.I.	1880-1954	philologist		1946
Tomsinskiy, S.G.	?	historian	1933	
Tomson, A.I.	1860-1935	linguist	1910	
Topchiyev, A.V.	1907-1962	organic chemist		1949
Toropov, N.A.	1906-1968	physico-chemist	1962	
Trakhtenberg, I.A.	1883-1960	economist		1939

Name	Born	Field	Year1	Year2
Trapeznikov, V.A.	1905–	automation and power eng	1953	1960
Traynin, A.N.	1883–1957	lawyer	1946	
Traynin, I.P.	1887–1949	lawyer		1939
Tret'yakov, P.N.	1909–	archeologist	1958	
Trever, K.V.	1892–	historian	1943	
Trofimuk, A.A.	1911–	geologist	1953	1958
Troshin, A.S.	1912–	cytologist	1960	
Trukhanovskiy, V.G.	1914–	historian	1964	
Trutnev, Yu.A.	1927–	physicist	1964	
Tselikov, A.I.	1904–	metallurgist	1953	1964
Tsereteli, G.V.	1904–	Arabist, Semitist	1946	1968
Tsitsin, N.V.	1898–	botanist		1939
Tsvetkov, V.N.	1910–	physicist	1968	
Tsytovich, N.A.	1900–	permafrost specialist	1943	
Tuchkevich, V.M.	1904–	physicist	1968	
Tudorovskiy, A.I.	1875–1964	physicist	1933	
Tugarinov, A.I.	1917–	geochemist	1966	
Tulaykov, N.M.	1875–1938	agronomist		1932
Tumanov, I.I.	1894–	physiologist	1953	
Tumanskiy, S.K.	1901–	mech eng	1964	1968

Name	Dates	Field		
Tupolev, A.N.	1888-	plane designer	1933	1953
Turayev, B.A.	1868-1920	historian		1918
Tyagunenko, V.L.	1920-	economist	1968	
Tyumenev, A.I.	1880-1959	historian		1932
Tyurin, I.V.	1892-1962	soil scientist	1946	1953
Udal'tsov, A.D.	1883-1958	historian	1939	
Ugolev, A.M.	1926-	physiologist	1966	
Ukhtomskiy, A.A.	1875-1942	physiologist	1932	1935
Urazov, G.G.	1884-1957	metallurgist	1939	1946
Ushakov, D.N.	1873-1942	linguist	1939	
Ushakov, S.N.	1893-1964	organic chemist	1943	
Usov, M.A.	1883-1939	geologist	1932	1939
Uspenskiy, F.I.	1845-1928	historian	1893	1900
Vallander, S.V.	1917-	mech eng	1966	
Val'ter, A.F.	?	mech eng	1933	
Val'ter, P.A.	?	technician	1933	
Vanichev, A.P.	1916-	power eng	1962	
Varentsov, M.I.	1902-	geologist	1953	
Varga, Ye.S.	1879-1964	economist		1939

Vavilov, N.I.	1887-1943	botanist, geneticist	1923	1929
Vavilov, S.I.	1891-1951	physicist	1931	1932
Vaynshteyn, B.K.	1921-	physicist	1962	
Vaynshteyn, L.A.	1920-	radio physicist	1966	
Vdovenko, V.M.	1907-	chemist	1958	
Vedeneyev, B.Ye.	1884-1946	hydraulic eng		1932
Vedenisov, B.N.	1869-1952	railroad transport specialist	1943	
Vekshinskiy, S.A.	1896-	electronics eng	1946	1953
Veksler, V.I.	1907-1966	physicist	1946	1958
Vekua, I.N.	1907-	mathematician	1946	1958
Velikanov, D.P.	1908-	automobile transport eng	1968	
Velikanov, M.A.	1879-1964	hydrologist	1939	
Velikhov, Ye.P.	1935-	physicist	1968	
Vel'yaminov, N.A.	1855-1920	surgeon		1913
Venediktov, A.V.	1887-1959	lawyer		1958
Vereshchagin, L.F.	1909-	physicist	1960	1966
Vernadskiy, V.I.	1863-1945	mineralogist	1908	1912

Vernov, S.N.	1910-	physicist	1953	1968
Veselovskiy, S.B.	1876-1952	historian		1946
Vesnin, V.A.	1882-1950	architect		1943
Veyts, V.I.	1905-1961	power eng	1933	
Vil'yams, V.R.	1863-1939	soil scientist		1931
Vinogradov, A.P.	1895-	biogeochemist	1943	1953
Vinogradov, I.M.	1891-	mathematician		1929
Vinogradov, P.G.	1854-1925	historian	1892	1914
Vinogradov, V.A.	1921-	economist	1966	
Vinogradov, V.V.	1895-1969	linguist		1946
Vinogradskiy, S.N.	1856-1953	microbiologist	1894		1923
Vinter, A.V.	1878-1958	power eng		1932
Vipper, R.Yu.	1859-1954	historian		1943
Vize, V.Yu.	1886-1954	physicist	1933	
Vladimirov, V.S.	1923-	mathematician	1968	
Vladimirskiy, V.V.	1915-	physicist	1962	
Vladimirtsov, B.Ya.	1884-1931	Mongolist	1923	1929
Vlasov, K.A.	1905-1964	geochemist, mineralogist	1953	

Vlasov, V.Z.	1906-1958	mech eng	1953	
Voyevodskiy, V.V.	1917-1967	physico-chemist	1958	1964
Vol'fkovich, S.I.	1896-	inorganic chemist	1946	
Volgin, V.P.	1879-1962	historian	1930	
Vol'kenshteyn, M.V.	1912-	chemist	1966	
Volobuyev, V.R.	1909-	soil scientist	1968	
Vologdin, A.G.	1896-	paleontologist	1939	
Vologdin, V.P.	1881-1953	technician	1939	
Vol'skiy, A.N.	1897-1966	metallurgist	1953	1960
Vonsovskiy, S.V.	1910-	physicist	1953	1966
Voronin, L.G.	1908-	physiologist	1968	
Vorozhtsov, N.N.	1907-	chemist	1958	1966
Voytsekhovskiy, B.V.	1922-	hydrodynamics specialist	1964	
Voznesenskiy, I.N.	1887-1946	mech eng	1939	
Vrevskiy, M.S.	1871-1929	physico-chemist	1929	
Vul, B.M.	1903-	physicist	1939	
Vul'f, G.V.	1863-1925	crystallographer	1921	
Vvedenskiy, B.A.	1893-1969	radio physicist	1934	1943

Vvedenskiy, N.Ye.	1852-1922	physiologist	1909	
Vyshinskiy, A.Ya.	1883-1954	lawyer		1939
Yablonskiy, S.V.	1924-	mathematician	1968	
Yachevskiy, A.A.	1863-1922	botanist	1923	
Yagich Vatroslav (I.V.)	1838-1923	philologist	1868	1881
Yakovkin, A.A.	1860-1936	chemist	1935	
Yakovlev, A.I.	1878-1951	historian	1929	
Yakovlev, A.S.	1906-	plane designer	1943	
Yakovlev, N.N.	1870-1966	paleontologist	1921	
Yakubovskiy, A.Yu.	1886-1953	Orientalist	1943	
Yanenko, N.N.	1921-	mech eng	1966	
Yangel', M.K.	1911-	mech eng		1966
Yanin, V.L.	1929-	historian	1966	
Yanovskiy, M.I.	1888-1949	technician	1943	
Yanshin, A.L.	1911-	geologist		1958
Yaroslavskiy, Ye.M.	1878-1943	historian		1939
Yartseva, V.N.	1906-	philologist	1968	
Yefimov, A.N.	1908-	economist	1964	
Yefimov, A.V.	1896-	historian	1939	

Yegolin, A.M.	1896–1959	lit historian	1946	
Yegorov, A.G.	1920–	philosopher	1962	
Yegorov, D.F.	1869–1931	mathematician	1924	1929
Yeliseyev, N.A.	1897–1966	geologist	1953	
Yelyutin, V.P.	1907–	metallurgist	1962	
Yemel'yanov, V.S.	1901–	metallurgist	1953	
Yenikolopov, N.S.	1924–	chemist	1966	
Yermakov, V.P.	1845–1922	mathematician	1884	
Yernshtedt, P.V.	1890–1966	philologist	1946	
Yudin, P.F.	1899–1968	philosopher	1939	1953
Yunusov, S.Yu.	1909–	chemist	1958	
Yur'yev, B.N.	1889–1957	aerodynamics specialist		1943
Zababakhin, Ye.I.	1917–	physicist	1958	1968
Zabolotnyy, D.K.	1866–1929	microbiologist		1929
Zalesskiy, M.D.	1877–1946	paleobotanist	1929	
Zaslavskaya, T.I.	1927–	economist	1968	
Zatsepin, G.T.	1917–	physicist	1968	
Zavalishin, D.A.	1900–1968	electr eng	1960	
Zavaritskiy, A.N.	1884–1952	geologist		1939

Zavarzin, A.A.	1886-1945	histologist		1943
Zavoyskiy, Ye.K.	1907-	physicist	1953	1964
Zaymovskiy, A.S.	1905-	metallurgist	1958	
Zefirov, A.P.	1907-	technol chemist	1968	
Zel'dovich, Ya.B.	1914-	physicist	1946	1958
Zelenin, D.K.	1878-1954	ethnographer	1925	
Zelenskiy, V.V.	1847-1918	zoologist	1893	1897
Zelinskiy, N.D.	1861-1953	organic chemist	1924	1929
Zemyatchenskiy, P.A.	1856-1942	soil scientist, geologist	1928	
Zenkevich, L.A.	1889-1970	oceanologist	1953	1968
Zernov, D.V.	1907-	electronics eng	1953	
Zernov, S.A.	1871-1945	hydrobiologist		1931
Zhavoronkov, N.M.	1907-	chemist	1953	1962
Zhebelev, S.A.	1867-1941	historian		1927
Zheltukhin, N.A.	1915-	mech eng	1968	
Zhemchuzhnikov, Yu.A.	1885-1957	geologist	1946	
Zhilin, P.A.	1913-	historian	1968	
Zhirmunskiy, V.M.	1891-	linguist	1939	1966

Zhuk, S.Ya.	1892-1957	hydraulic eng		1953
Zhukov, A.B.	1901-	biologist		1966
Zhukov, B.P.	1912-	technol chemist	1968	
Zhukov, I.I.	1880-1949	chemist	1946	
Zhukov, M.F.	1917-	aerodynamics specialist	1968	
Zhukov, Ye.M.	1907-	historian	1946	1958
Zhukovskiy, N.Ye.	1847-1921	hydromechanic	1894	
Zhurkov, S.N.	1905-	physicist	1958	1968
Zlatogorov, S.I.	1873-1931	microbiologist	1929	
Zutis, Ya.Ya.	1893-1962	historian	1953	
Zverev, M.S.	1903-	astronomer	1953	
Zvonkov, V.V.	1891-1965	transport eng	1939	

Members of the Russian Academy of Sciences (1725-1917)

ABIKH (ABICH), Wilhelm-Hermann (German Vasil'yevich), academician ordinarius for oryctognosy and mineralogical chemistry from 8 Jan 1853; left 22 Dec 1865 (hon member from 14 Jan 1866); Born: 11 Dec 1806 in Berlin; Died: 1 July 1886 in Vienna.

ADADUROV (ADODUROV), Vasiliy Yevdokimovich, junior associate for higher mathematics from 26 Oct 1733; left in Apr 1741 (hon member from 1778); Born: 15 Mar 1709 in Novgorod; Died: 5 Nov 1780 in Petersburg.

ADAMS, Michael-Friedrich (Mikhail Ivanovich), junior associate for zoology from 27 Mar 1805; left 11 Mar 1809 (corresp member from 1 Feb 1804; hon member from 1814); Born: 1780 in Moscow; Died: prior to 1829.

ALEKSEYEV, Nikolay Nikolayevich, junior associate for mathematics from 5 Oct 1879; Born: 5 May 1829; Died: 2 Mar 1881 in Petersburg.

AMMAN, Johann, Dr of Med; prof of botany and natural history from 2 June 1733; Born: 1707 in Schaffhausen; Died: 4 Dec 1741 in Petersburg.

ANDUSOV, Nikolay Ivanovich, academician ordinarius for geognosy and paleontology from 3 May 1914 (corresp member from 29 Dec 1910); Born: 7 Dec 1861 in Odessa; Died: 27 Apr 1924 in Prague.

* ANUCHIN, Dmitriy Nikolayevich, academician ordinarius for zoology from 10 Feb 1896 (hon member from 18 Apr 1898); left 30 May 1898.

ARSEN'YEV, Konstantin Ivanovich, academician ordinarius, Dept of Russian Language and Lit from 19 Oct 1841 (corresp member, Political Economy Dept from 29 Dec 1826; member, Russian Academy from 14 Mar 1836); Born: 12 Oct 1789 in vil Marakhanovo, Kostroma Province; Died: 29 Nov 1865 in Petrozavodsk.

---

* For academicians marked with an asterisk see lists of members and corresponding members of the USSR Academy of Sciences.

BAKLUND, Oskar Andreyevich, academician ordinarius for astronomy from 3 Dec 1883 (corresp member from 29 Dec 1881); Born: 16 Apr 1846 in Lengem, Sweden; Died: 16 Aug 1916 in Pulkovo.

* BARTOL'D, Vasiliy Vladimirovich, academician ordinarius for lit and history of Asian peoples from 12 Oct 1913.

BAYYER (BAYER), Gottlieb-Siegfried, prof, Chair of Greco-Roman Antiquities from 3 Dec 1725; left late 1737; Born: 6 Jan 1694 in Koenigsberg; Died: 10 (21) Feb 1738.

BEKENSHTEYN (BECKENSTEIN), Johann-Simon, prof of jurisprudence from 3 Dec 1725 to May 1735 (hon member from 1738); Born: in Danzig; Died: prior to 1744.

BEKETOV, Nikolay Nikolayevich, academician ordinarius for general chemistry from 13 Dec 1886 (corresp member for physical sciences from 29 Dec 1877); Born: 1 Jan 1827 on Beketovka Estate, Penza Province; Died: 30 Nov 1911 in Petersburg.

* BELOPOL'SKIY, Aristarkh Apollonovich, junior associate for astronomy from 13 May 1900.

BER (von BAER), Karl-Ernst (Karl Maksimovich), academician ordinarius for zoology from 15 Aug 1828 (corresp member from 26 Dec 1826); left 23 Oct 1830 (hon member from 1830); rejoined 11 Apr 1834; from 1846 academician ordinarius for anatomy and physiology; left again 27 Oct 1862 (hon member from 2 Nov 1862); Born: 17 Feb 1792 on Pieg Estate, Estland; Died: 16 Nov 1876 in Derpt.

BEREDNIKOV, Yakov Ivanovich, junior associate, Dept of Russian Language and Lit from 19 Oct 1841; academician extraordinarius from 3 May 1845; academician ordinarius from 6 Feb 1847; Born: 7 Oct 1793 in Petersburg; Died: 28 Sept 1854 in Petersburg.

BERNULLI (BERNOULLI), Daniel, second son of Jacob Bernoulli; prof of physiology from 5 July 1725; prof of mathematics from 1727; academician from 27 Oct 1725 to 24 June 1733 (hon member from 1733); Born: 29 Jan 1700 in Groeningen; Died: 17 Mar 1782 in Basel.

BERNULLI (BERNOULLI), Jacob, youngest son of Prof Johann Bernoulli; junior associate for mathematics from 20 Apr 1786; academician ordinarius from 27 Sept 1787; Born: 17 Oct 1759 in Basel; Died: 3 July 1789.

BERNULLI (BERNOULLI), Nicholas II, son of Johann Bernoulli I; prof of mathematics from 1725; academician from 27 Oct 1725; Born: 27 Jan 1695 in Basel; Died: 29 July 1726 in Petersburg.

BESTUZHEV-RYUMIN, Konstantin Nikolayevich, academician ordinarius, Dept of Russian Language and Lit from 3 Mar 1890 (corresp member from 29 Dec 1892); Born: 14 May 1829 in vil Kudreshki, Nizhniy Novgorod Province; Died: 2 Jan 1897 in Petersburg.

BETLING (BOHTLINGK), Otto (Otton Nikolayevich), junior associate for Sanskrit from 5 Mar 1842; academician extraordinarius from 13 Dec 1845; academician ordinarius from 4 Aug 1855; left 11 June 1894 (hon member from 3 Sept 1894); Born: 30 May 1815 in Petersburg; Died: 1 Apr 1904 in Leipzig.

BEL'SHTEYN, Fyodor Fyodorovich, academician ordinarius for technology and chemistry applied to arts and crafts from 13 Dec 1886 (corresp member for physical sciences from 29 Dec 1883); Born: 5 Feb 1838 in Petersburg; Died: 5 Oct 1906 in Petersburg.

BEZOBRAZOV, Vladimir Pavlovich, junior associate for political economy and statistics from 4 Dec 1864; academician extraordinarius from 4 Aug 1867; Born: 3 Jan 1829 in Vladimir; Died: 30 Aug 1889 in Dmitrovo, Moscow Province.

BILYARSKIY, Pyotr Spiridonovich, junior associate, Dept of Russian Language and Lit from 3 June 1860; academician extraordinarius from 11 Jan 1863; Born: 19 June 1815 in Bilyar, Kazan' Province; Died: 2 Jan 1867 in Odessa.

BONGARD, Heinrich-Gustav (Gustav Petrovich), junior associate for botany from 5 May 1830; academician extraordinarius from 20 May 1836 (corresp member from 1 Apr 1829); Born: 12 Sept 1786 in Bonn; Died: 25 Aug 1839 in Petersburg.

BORISOV, Ivan Alekseyevich (Bishop Innokentiy), academician ordinarius, Dept of Russian Language and Lit from 19 Oct 1841 (member, Russian Academy from 11 Jan 1836); Born: 1800 in Sevsk, Oryol Province; Died: 26 May 1857 in Odessa.

* BORODIN, Ivan Parfen'yevich

BRANDT, Johann-Friedrich (Fyodor Fyodorovich), junior associate for zoology from 15 Dec 1830; academician extraordinarius from 16 May 1832; academician ordinarius from 14 June 1833; Born: 25 May 1802 in Jüterbog, Prussian Saxony; Died: 3 July 1879 in Merrekyul, near Narva.

BRAUN, Joseph (Johann ?)-Adam, prof of philosophy from 23 Jan 1748; Born: 1712 in Asch; Died: 22 Sept 1768 in Petersburg.

BREDIKHIN, Fyodor Aleksandrovich, academician ordinarius for astronomy from 17 Mar 1890 (corresp member from 29 Dec 1877); Born: 26 Nov 1831 in Nikolayev, Kherson Province; Died: 1 May 1904 in Petersburg.

BREME (BREHME), Johann-Friedrich, junior associate for history from 1 Sept 1737 to 1 Aug 1747; Born: in Revel.

BROSSE (BROSSET), Marie-Felicité (Mariy Ivanovich), junior associate for Armenian and Georgian lit from 2 Dec 1836; academician extraordinarius from 2 Mar 1838; academician ordinarius from 4 Dec 1847; Born: 24 Jan 1802 in Paris; Died: 3 Sept 1880 in Chatelherault, France.

BUKSBAUM (BUXBAUM), Johann-Christian, prof of botany until Feb 1727; left 4 Aug 1729; Born: Oct 1694 in Merseburg, Saxony; Died: 7 July 1730 at Wermsdorf, near Merseburg.

BULGAKOV, Mikhail Petrovich (Bishop Makariy), academician ordinarius, Dept of Russian Language and Lit from 4 Nov 1854; Born: 19 Sept 1816; Died: 9 June 1882 in Moscow.

BUNGE, Nikolay Khristianovich, academician ordinarius for political economy and financial science from 3 Mar 1890 (corresp member from 29 Dec 1859; hon member from 29 Dec 1881); Born: 11 Nov 1823 in Kiev; Died: 3 June 1895 in Tsarskoye Selo.

BUNYAKOVSKIY, Viktor Yakovlevich, junior associate for pure mathematics from 7 May 1828; academician extraordinarius from 24 Mar 1830; academician ordinarius from 8 Jan 1841; Born: 3 Dec 1804 in Bar, Podol'sk Province; Died: 30 Nov 1889 in Petersburg.

BUSLAYEV, Fyodor Ivanovich, academician ordinarius, Dept of Russian Language and Lit from 3 June 1860 (corresp member from 29 Dec 1852); Born: 13 Apr 1818 in Kerensk, Penza Province; Died: 31 July 1897 in Lublin, Moscow Province.

BUSSE, Ivan Fomich (Johann-Heinrich), junior associate for history from 29 Jan 1795 to 1800 (hon member from 1800); Born: 14 Sept 1763 in Gardelegen, Brandenburg; Died: 20 July 1835 in Grabow, Stettin.

BUTKOV, Pyotr Grigor'yevich, academician ordinarius, Dept of Russian Language and Lit from 19 Oct 1841 (member, Russian Academy from 6 Feb 1837); Born: 17 Dec 1775; Died: 12 Dec 1857 in Petersburg.

BUTLEROV, Aleksandr Mikhaylovich, junior associate for chemistry from 6 Mar 1870; academician extraordinarius from 3 Dec 1871; academician ordinarius from 18 Dec 1874; Born: 25 Aug 1828 in Chistopol'; Died: 5 Aug 1886 in vil Butlerovka, Kazan' Province.

BYCHKOV, Afanasiy Fyodorovich, academician extraordinarius, Dept of Russian Language and Lit from 14 Jan 1866; academician ordinarius from 5 Dec 1869 (corresp member from 29 Dec 1855); Born: 15 Dec 1818 in Friedrichsham; Died: 2 Apr 1899 in Petersburg.

BYUL'FINGER (BÜLFFINGER), Georg-Bernhard, prof of logic and metaphysics from 1 Mar 1725; prof of experimental and theoretical physics from 1726; member until 1730 (hon member from 1731); Born: 23 Jan 1693 in Cannstatt; Died: 18 Feb 1750 in Stuttgart.

BYURGER (BÜRGER), Michael, Dr of Med; prof of chemistry and practical medicine from Sept 1725; Born: in Memel; Died: 22 July 1726 in Petersburg.

CHEBYSHYOV, Pafnutiy L'vovich, junior associate for applied mathematics from 14 May 1853; academician extraordinarius from 3 Aug 1856; academician ordinarius from 6 Feb 1859; Born: 14 May 1821 in vil Okatovo, Kaluga Province; Died: 26 Nov 1894 in Petersburg.

CHERNYSHYOV, Feodosiy Nikolayevich, junior associate for geology from 11 Jan 1897; academician extraordinarius from 4 Dec 1899; academician ordinarius for geology and paleontology from 2 May 1909; Born: 12 Sept 1856 in Kiev; Died: 2 Jan 1914 in Petersburg.

CHYORNYY, Fyodor, junior associate for astronomy from 31 Dec 1785; Died: 25 July 1790.

DASHKEVICH, Nikolay Pavlovich, academician ordinarius, Dept of Russian Language and Lit from 7 Apr 1907 (corresp member from 29 Dec 1902); Born: 4 Aug 1852 in vil Bezhevo, Volhynian Province; Died: 20 Jan 1908 in Kiev.

DAVYDOV, Ivan Ivanovich, academician ordinarius, Dept of Russian Language and Lit from 19 Oct 1841; Born: 15 June 1794 in Tver'; Died: 15 Nov 1863 in Moscow.

DELIL' (DE L'ISLE), Joseph-Nicole (Osip Nikolayevich), prof of astronomy from 8 July 1725 to 23 Jan 1747 (hon member from 1747); Born: 4 Apr 1688 in Paris; Died: 12 Sept 1768 in Paris.

DELIL' DE LA KROYYER (DE L'ISLE DE LA CROYÈRE), Louis, brother of Joseph-Nicole de l'Isle; prof extraordinarius of astronomy from early 1727; Died: 10 Oct 1741 in Kamchatka.

DORN, Johann-Albrecht-Bernhard (Boris Andreyevich), junior associate for Oriental languages from 1 Feb 1839; academician extraordinarius from 4 June 1842; academician ordinarius from 6 Mar 1852 (corresp member from 29 Dec 1835); Born: 11 May 1805 in Scheuerfeld, near Coburg; Died: 19 May 1881 in Petersburg.

DROZDOV, Vasiliy Mikhaylovich (Metropolitan Filaret), academician ordinarius, Dept of Russian Language and Lit from 19 Oct 1841 (hon member from 29 Dec 1827; member, Russian Academy from 5 Oct 1818); Born: 26 Dec 1782 in Kolomna; Died: 19 Nov 1867 in Moscow.

DUBROVICH, Nikolay Fyodorovich, junior associate for history and Russian antiquities from 7 Mar 1887; academician extraordinarius from 1 Sept 1890; academician ordinarius from 4 Dec 1899; (corresp member for historical and political sciences from 29 Dec 1877); Born: 26 Nov 1837 in vil Korytovo, Pskov Province; Died: 12 June 1904 in Petersburg.

DUBROVSKIY, Pyotr Pavlovich, junior associate, Dept of Russian Language and Lit from 2 June 1855; academician extraordinarius from 26 Feb 1858; left 8 June 1862 (corresp member, Dept of Russian Language and Lit from 29 Dec 1853 and from 29 Dec 1862); Born: 14 June 1812 in Chernigov; Died: 29 Oct 1882 in Skernevitsy.

* D'YAKONOV, Mikhail Aleksandrovich, junior associate for history and Russian antiquities from 3 Sept 1905.

DYUVERNUA (DU VERNOI), Johann-Georg, Dr of Med; prof of anatomy, surgery and zoology from 3 Nov 1725; left 16 June 1741; Born: 1691 in Mummelgard; Died: 1759 in Amstetten, Württemberg.

EPINUS (AEPINUS), Franz-Ulrich-Theodor, prof of physics from Oct 1756; left in May 1798; Born: 13 Dec 1724 in Rostock; Died: 10 Aug 1802 in Derpt.

ERNSHTEDT, Viktor Karlovich, junior associate for classical philology and archeology from 1 May 1893; academician extraordinarius from 5 Dec 1898; Born: 14 Dec 1854 in Petersburg; Died: 21 Aug 1902 in Petersburg.

EYLER (EULER), Johann-Albrecht, son of Leonard Euler; prof of physics from 17 July 1766; conference secr from 22 Feb 1769; Born: 16 (27) Nov 1734 in Petersburg; Died: 6 Sept 1800 in Petersburg.

EYLER (EULER), Leonard, junior associate for physiology from 1726; prof of physics from 1731; prof of higher mathematics from 1733; left 5 June 1741; returned to Russia 17 June 1766 (hon member from 1742 to 1766); Born: 4 Apr 1707 in Basel; Died: 7 Sept 1783 in Petersburg.

* FAMINTSYN, Andrey Sergeyevich, junior associate for botany from 1 Dec 1878.

FERBER, Johann-Jacob, academician ordinarius for mineralogy from 24 Apr 1783; left 1786 (hon member from 1785); Born: 29 Aug 1748 in Karlskron; Died: 17 Apr 1790 in Bern.

FISHER (FISCHER), Johann-Eberhard, junior associate from 1732; prof of history and antiquities from 28 Nov 1747; Born: 10 Jan 1697 in Esslingen, Württemberg; Died: 13 Sept 1771 in Petersburg.

FLEYSHER (FLEISCHER), Heinrich-Leberecht, junior associate from 29 May 1835 (corresp member from 29 Dec 1849); Born: 1801; Died: 10 Feb 1888.

FORTUNATOV, Filipp Fyodorovich, academician ordinarius, Dept of Russian Language and Lit from 7 Mar 1898 (corresp member from 29 Dec 1895); Born: 2 Jan 1848 in Vologda Province; Died: 20 Sept 1914 in Kosolma, Olonets Province.

FREN (FRAEHN), Christian-Martin (Khristian Danilovich), academician ordinarius for Oriental antiquities from 24 Sept 1817; Born: 12 May 1782 in Rostock; Died: 16 Aug 1851 in Petersburg.

FRITSSHE (FRITZSCHE), Karl-Julius (Yuliy Fyodorovich), junior associate for chemistry from 24 Aug 1838; academician extraordinarius from 6 Apr 1844; academician ordinarius from 10 Apr 1852; Born: 29 Oct 1808 in Neustadt, near Stolpen, Saxony; Died: 8 June 1871 in Dresden.

FUS (FUSS), Nikolay Ivanovich, junior associate from 15 Jan 1776; academician ordinarius for higher mathematics from 13 Feb 1783; secr from 17 Sept 1800; Born: 18 (30) Jan 1755 in Basel; Died: 23 Dec 1825 in Petersburg.

FUS (FUSS), Paul-Heinrich (Pavel Nikolayevich), junior associate for pure mathematics from 10 June 1818; academician extraordinarius from 29 Jan 1823; academician ordinarius from 15 Feb 1826; Born: 21 May 1798 in Petersburg; Died: 10 Jan 1855 in Petersburg.

* FYODOROV, Yevgraf Stepanovich, junior associate for mineralogy from 5 May 1901; left 28 May 1905; reelected 1 Feb 1919.

FYODOROVICH, Georg-Friedrich, prof of jurisprudence from 22 Feb 1760; left 1 Mar 1770; Born: in Brandenburgian Prussia.

GADOLIN, Aksel' Vil'gel'movich, academician extraordinarius for physics from 5 Dec 1875; academician ordinarius from 1 Dec 1890 (corresp member from 29 Dec 1873); Born: 12 June 1828 in Finland; Died: 15 Dec 1892 in Petersburg.

GAKMAN (HACKMANN), Johann-Friedrich, junior associate for history from 22 Aug 1782; left in late 1784 (hon member from 1784).

GAMEL' (HAMEL), Iosif Khristianovich, academician ordinarius for technology and chemistry applied to arts and crafts from 4 Mar 1829 (corresp member from 23 June 1813); Born: 30 Jan 1788 in Sarept; Died: 22 Sept 1862 in London.

GANRI, Moris (Maurice HENRY), academician ordinarius for astronomy from 7 July 1796; left 15 June 1800 (hon member from 1795 to 1796 and from 1800 to 1825); Born: 30 May 1763 in Sauvigny, near Toul; Died: 1825 in Paris.

GEBENSHTREYT (HEBENSTREIT), Johann-Christian, prof of botany from 11 Aug 1749; left for Leipzig in 1753; rejoined 21 Jan 1756; left again in May 1759 (hon member from 1754); Born: 1720 near Naumburg; Died: 27 Sept 1775 in Leipzig.

GELLERT, Christlieb-Ehregott, junior associate for chemistry from 25 June 1736; left in July 1744; Born: 11 Aug 1711 in Hainichen, Saxony; Died: 18 May 1795 in Freiberg.

GEL'MERSEN (HELMERSEN), Grigoriy Petrovich, junior associate for geognosy and paleontology from 3 Feb 1844; academician extraordinarius from 5 June 1847; academician ordinarius from 2 Mar 1850; Born: 29 Sept 1903 in Duckershof, near Derpt; Died: 3 Feb 1885 in Petersburg.

GEORGI, Johann-Gottlieb (Ivan Ivanovich), junior associate for chemistry from 15 Jan 1776; academician ordinarius for chemistry from 13 Feb 1783; Born: 31 Dec 1729 in Wachholzhagen, Pomerania; Died: 27 Oct 1802 in Petersburg.

GERMAN (HERRMANN), Franz-Johann-Benedikt (Ivan Filippovich), academician ordinarius for mineralogy from 11 Feb 1790; from 2 Dec 1801 head, Yekaterinburg Mining

Board (hon member of Academy from 1786 to 1815); Born: 14 Mar 1755 in Mariahof, Styria; Died: 31 Jan 1815 in Petersburg.

GERMAN (HERMANN), Jacob, prof of higher mathematics from 8 Jan 1725; left 18 Nov 1730 (hon member from 1731); Born: 16 July 1678 in Basel; Died: 14 July 1733 in Basel.

GERMAN (HERMANN), Karl-Theodor (Karl Fyodorovich), junior associate for statistics and political economy from 27 Mar 1805; academician extraordinarius from 17 Jan 1810; academician ordinarius from 11 Dec 1835; Born: 5 Sept 1767 in Danzig; Died: 19 Dec 1838 in Petersburg.

GERTNER (GÄRTNER), Joseph, prof of botany from 1768; left in 1770 (hon member from 1770); Born: 11 Mar 1732 in Calw, Württemberg; Died: 2 (13) July 1791 in Calw.

GESS (HESS), Hermann-Heinrich (German Ivanovich), junior associate for general chemistry from 29 Oct 1828; academician extraordinarius from 11 Aug 1830; academician ordinarius from 14 May 1834; Born: 7 Aug 1802 in Geneva; Died: 30 Nov 1850 in Petersburg.

GEYNZIUS (HEINSIUS), Gottfried, prof extraordinarius of astronomy from 20 May 1736; left 22 May 1744 (hon member from 1744); Born: Apr 1709 in Naumburg; Died: 21 May 1769 in Leipzig.

GIL'DENSHTEDT (GÜLDENSTADT), Johann-Anton, Dr of Med; junior associate from 2 Oct 1769; prof of natural history from 8 Apr 1771; Born: 29 Apr 1745 in Riga; Died: 23 Mar 1781 in Petersburg.

GMELIN, Johann-Georg, junior associate from 1727; arrived in Petersburg 19 Aug 1727; prof of chemistry and natural history from 22 Jan 1731; left in 1747 (1748 ?); Born: 12 Aug 1709 in Tübingen; Died: 20 May 1755 in Tübingen.

GMELIN, Samuel-Gottlieb, prof of botany from 4 Apr 1767; Born: 23 July 1745 in Tübingen; Died: 27 July 1774 in Derbent.

GOL'DBAKH (GOLDBACH), Christian, prof of mathematics from 1 Sept 1725; left 18 Mar 1742 (hon member from

1742); member, Foreign Collegium from 1747; first conference secr and councillor of the Academy from 1725 to 1728; Born: 8 Mar 1690 in Koenigsberg; Died: 20 Nov 1764 in Koenigsberg.

GOLITSYN, Boris Borisovich, junior associate for physics from 4 Dec 1893; academician extraordinarius from 5 Dec 1898; academician ordinarius from 5 Apr 1908; Born: 18 Feb 1862 in Petersburg; Died: 4 May 1916 in Novyy Peterhof.

GOLOVIN, Mikhail Yevsev'yevich, junior associate for mathematics from 15 Jan 1776; left 19 Jan 1786 (hon member from 1786); Born: 1756 in Arkhangel'sk Province; Died: 8 June 1790 in Petersburg.

GOLUBINSKIY, Yevgeniy Yevstigneyevich, academician ordinarius, Dept of Russian Language and Lit from 19 Apr 1903 (corresp member from 29 Dec 1882); Born: 28 Feb 1834 in Kostroma Province; Died: 7 Jan 1912 in Moscow.

GORNER (HORNER), Johann-Kaspar, junior associate for astronomy from 16 Oct 1806; left 21 Sept 1808 (corresp member from 21 Sept 1808); Born: 21 Mar 1774 in Zurich; Died: 3 Nov 1834 in Zurich.

GREFE (GRÄFE), Christian-Friedrich (Fyodor Bogdanovich), academician ordinarius for Greco-Roman lit from 8 Mar 1820 (corresp member from 16 Dec 1818); Born: 2 July 1780 in Chemnitz; Died: 30 Nov 1851 in Petersburg.

GRISHOV (GRISCHOW), Augustin-Nathaniel, prof of astronomy from 15 Feb 1751; Born: 29 Sept 1726 in Berlin; Died: 4 June 1760 in Petersburg.

GROSS, Christopher-Friedrich, junior associate from 1725; prof extraordinarius of moral philosophy from 29 Jan 1726; arrived in Petersburg in the fall of 1725; left in 1731 (then councillor, Braunschweig Embassy; hon member of Academy from 1731); Born: in Württemberg; Died: 2 Jan 1742 in Petersburg.

GROT, Yakov Karlovich, junior associate, Dept of Russian Language and Lit from 2 June 1855; academician extraordinarius from 22 Dec 1856; academician ordinarius from 26 Feb 1858 (corresp member from 29 Dec 1852); Born: 15 Dec 1812 in Petersburg; Died: 24 May 1893 in Petersburg.

GUR'YEV, Semyon Yemel'yanovich, junior associate for mathematics from 26 May 1796; academician ordinarius from 31 Jan 1798; Born: 1764 (?); Died: 11 Dec 1813 in Petersburg.

* IKONNIKOV, Vladimir Stepanovich, academician ordinarius, Dept of Russian Language and Lit (History) from 8 Feb 1914.

IMSHENETSKIY, Vasiliy Grigor'yevich, academician ordinarius for pure mathematics from 4 Dec 1881; Born: 4 Jan 1832 at Izhevskiy Zavod, Vyatka Province; Died: 24 May 1892 in Moscow.

INOKHODTSEV, Pyotr Borisovich, junior associate for astronomy from 10 Oct 1768; prof extraordinarius of astronomy from early 1779; prof ordinarius from 10 Mar 1783; 16 Feb 1797 appointed censor in Riga, but remained a member of the Academy; 23 July 1799 rejoined the Academy; Born: 21 Nov 1742 in Moscow; Died: 27 Oct 1806 in Petersburg.

* IPAT'YEV, Vladimir Nikolayevich, Born: 9 Nov 1867 in Moscow; left Academy; Died: 1952 abroad.

ISLEN'YEV, Ivan Ivanovich, junior associate, Geographical Dept, Academy of Sci from 1771; Died: 22 Feb 1784.

* ISTRIN, Vasiliy Mikhaylovich

KAAU-BURGAV (KAAU-BOERHAAVE), Abraham, prof of anatomy and physiology from Dec 1747; Born: 5 Jan 1715 in The Hague; Died: 14 July 1758 in Petersburg.

KACHENOVSKIY, Mikhail Trofimovich, academician ordinarius, Dept of Russian Language and Lit from 19 Oct 1841 (corresp member from 29 Dec 1832; member, Russian Academy from 8 Mar 1819); Born: 1 Nov 1775 in Khar'kov; Died: 19 Apr 1842 in Moscow.

KALACHOV, Nikolay Vasil'yevich, academician ordinarius for history and Russian antiquities from 2 Apr 1883 (corresp member from 29 Dec 1858); Born: 26 May 1819 in vil Aleksin, Vladimir Province; Died: 25 Oct 1885 on Volkhonshina Estate, Saratov Province.

* KARPINSKIY, Aleksandr Petrovich, junior associate for geology from 7 Feb 1886.

* KARSKIY, Yevfimiy Fyodorovich

KEMTS (KÄMTZ), Ludwig-Friedrich (Lyudvig Martinovich), academician ordinarius for physics from 5 Nov 1865; Born: 11 Jan 1801 in Treptow, Pomerania; Died: 8 Dec 1867 in Petersburg.

KEPPEN (KÖPPEN), Pyotr Ivanovich, junior associate for statistics from 27 Jan 1837; academician extraordinarius from 20 Dec 1839; academician ordinarius from 1 Apr 1843 (corresp member from 26 Dec 1826); Born: 19 Feb 1793 in Khar'kov; Died: 23 May 1864 in Karabakh, Tavrida Province.

KIRKHGOF (KIRCHHOF), Konstantin-Gottlieb-Sigismund (Konstantin Sigizmundovich), junior associate for chemistry from 8 Nov 1809; academician extraordinarius from 1 Apr 1812; left 4 Mar 1818 (corresp member from 4 Nov 1807); Born: 4 Feb 1764 in Teterow, Mecklenburg-Schwerin; Died: 15 Feb 1833 in Petersburg.

KLAPROT (KLAPROTH), Heinrich-Julius, junior associate for Oriental languages and lit from 1 Sept 1804; academician extraordinarius from 11 Mar 1807; dismissed 15 May 1817; Born: 11 Oct 1783 in Berlin; Died: 8 Aug 1835 in Paris.

KLEYNFEL'D (KLEINFELD), Martin, junior associate for anatomy from 11 May 1748; Born: in Rostock; Died: 29 Jan 1761 in Petersburg.

KLYUCHEVSKIY, Vasiliy Osipovich, academician ordinarius for history and Russian antiquities from 7 Oct 1900 (corresp member from 29 Dec 1889); Born: 16 Jan 1841 in vil Voskresenskoye, Penza Province; Died: 12 May 1911 in Moscow.

KOCHETOV, Ioakim Semyonovich, academician ordinarius, Dept of Russian Language and Lit from 7 Mar 1846 (hon member from 21 Nov 1841); Born: 1 Sept 1787; Died: 16 Mar 1854 in Petersburg.

* KOKOVTSOV, Pavel Konstantinovich, junior associate for lit and history of Asian peoples from 19 Apr 1903.

KOKSHAROV, Nikolay Ivanovich, junior associate for crystallographical ornithognosis from 2 June 1855; academician extraordinarius from 24 May 1858; academician ordinarius from 4 Mar 1866; Born: 23 Nov 1818 near Ust'-Kamenogorsk, Tomsk Province; Died: 21 Dec 1892 in Petersburg.

KOL' (KOHL), Johann-Christopher, prof of rhetoric and church history from 7 Feb 1725; left in Aug 1727; Born: 10 Mar 1698 in Kiel; Died: 9 Oct 1778.

KOLLINS (COLLINS), Eduard-Albert-Christopher-Ludwig (Eduard Davydovich), junior associate for mathematics from 26 Jan 1814; academician extraordinarius from 26 Jan 1820; academician ordinarius from 15 Feb 1826; Born: 3 July 1791 in Petersburg; Died: 4 Aug 1840 in Petersburg.

* KONDAKOV, Nikodim Pavlovich, academician ordinarius for classical philology and archeology from 23 May 1898; academician ordinarius, Dept of Russian Language and Lit from 5 Feb 1900 (corresp member from 29 Dec 1892); Born: 1 Nov 1844 in Novooskol Uyezd, Kursk Province; Died: 17 Feb 1925 in Prague.

KONONOV, Aleksey Kononovich, junior associate for physics from 25 June 1789; academician extraordinarius from 26 Jan 1795; Born: 1766; Died: 5 Oct 1795 in Petersburg.

KORKUNOV, Mikhail Andreyevich, junior associate, Dept of Russian Language and Lit from 2 Oct 1847; academician extraordinarius from 1 Nov 1851; academician ordinarius from 2 Aug 1857; Born: 2 Sept 1806 in Penza; Died: 13 Jan 1858 in Petersburg.

KORSH, Fyodor Yevgen'yevich, academician ordinarius, Dept of Russian Language and Lit from 15 Jan 1900 (corresp member from 29 Dec 1893); Born: 22 Apr 1843 in Moscow; Died: 16 Feb 1915 in Moscow.

KORZHINSKIY, Sergey Ivanovich, junior associate for botany from 9 Jan 1893; academician extraordinarius from 7 Dec 1896; Born: 26 Aug 1861 in Astrakhan'; Died: 18 Nov 1900 in Petersburg.

KOTEL'NIKOV, Semyon Kirillovich, junior associate from 22 Mar 1751; prof extraordinarius of mathematics from Jan 1757; prof of higher mathematics from 9 Oct 1760; left on 16 Feb 1797 with his appointment as public censor (hon member from 1797); member, Russian Academy from 1783; Born: 1723; Died: 1 Apr 1806 in Petersburg.

* KOTLYAREVSKIY, Nestor Aleksandrovich

KOVALEVSKIY, Aleksandr Onufriyevich, academician ordinarius for zoology from 24 Mar 1890 (corresp member from 29 Dec 1893); Born: 7 Nov 1840 on Volkovo Estate, Vitebsk Province; Died: 9 Nov 1901 in Petersburg.

KOVALEVSKIY, Iosif Mikhaylovich, academician ordinarius for Tibetan and Mongolian lit from 4 Dec 1847; never became a full member of the Academy (corresp member from 29 Dec 1837); Born: 1800 in Grodno; Died: 26 Oct 1878 in Warsaw.

KOVALEVSKIY, Maksim Maksimovich, academician ordinarius for state law from 29 Mar 1914 (corresp member from 29 Dec 1899); Born: 27 Aug 1851 in Khar'kov; Died: 23 Mar 1916 in Petrograd.

KOZITSKIY, Grigoriy Vasil'yevich, junior associate from 15 Mar 1759 (hon member from 1767); Born: 1724; Died: 26 Dec 1775 in Moscow.

KRAFT (KRAFFT), Georg-Wolfgang, junior associate from 1727; prof of mathematics from 1 Jan 1731; prof of physics from 31 Jan 1731; secr from 1730 to 1733; left 29 May 1744 (hon member from 1744); Born: 15 July 1701 in Tuttlingen, Württemberg; Died: 16 July 1754 in Tübingen.

KRAFT (KRAFFT), Wolfgang-Ludwig (Login Yur'yevich), son of Georg-Wolfgang Krafft; junior associate from 22 Dec 1768; prof of experimental physics from 8 Apr 1781; Born: 25 Aug 1743 in Petersburg; Died: 20 Nov 1814 in Petersburg.

KRAMER (CRAMER), Adolph-Bernhard, junior associate for history from 17 Oct 1732; Born: in Herford, Westphalia; Died: 20 Nov 1734 in Petersburg.

KRASHENINNIKOV, Stepan Petrovich, junior associate for natural history from 25 July 1745; prof of botany and natural history from 11 Apr 1750; Born: 1713 (or 1711) in Moscow; Died: 25 Feb 1755 in Petersburg.

KRASIL'NIKOV, Andrey Dmitriyevich, junior associate for astronomy from May 1753; Born: 1705; Died: 15 Feb 1773 in Petersburg.

KRATTSENSHTEYN (KRATZENSTEIN), Christian - Gottlieb, prof of mechanics from 22 Mar 1748; left in Aug 1753 (hon member from 1753); Born: 30 Jan 1723 in Vernigorod; Died: 6 July 1795 in Copenhagen.

KREL' (KREHL), Christopher-Ludolph-Ehrenfried, junior associate for Islamic languages from 2 June 1855, although his appointment was not approved by Minister of Education A.S. Norov; Born: 1825; Died: 1901.

KRUG, Johann-Philipp (Filipp Ivanovich), junior associate for history from 27 Mar 1805; academician extraordinarius from 11 Mar 1807; academician ordinarius from 16 Aug 1815; Born: 29 Jan 1764 in Halle; Died: 4 June 1844 in Petersburg.

KRUZIUS (CRUSIUS), Christian, junior associate from 28 Mar 1740; prof of antiquities and the history of letters from 6 Oct 1746; left 1749-1750; Born: 1715 in Wollbach, Saxony; Died: 7 Feb 1767 in Wittenberg.

* KORNAKOV, Nikolay Semyonovich

* KRYLOV, Aleksey Nikolayevich

KRYLOV, Ivan Andreyevich, academician ordinarius, Dept of Russian Language and Lit from 19 Oct 1841 (member, Russian Academy from 16 Dec 1811); Born: 9 Feb 1768 in Moscow; Died: 9 Nov 1844 in Petersburg.

KUNIK, Ernst-Eduard (Arist Aristovich), junior associate for history from 5 Oct 1844; academician extraordinarius from 2 Mar 1850; Born: 14 Oct 1814 in vil Gramovitsy, near Jauer, Prussian Silesia; Died: 18 Jan 1899 in Petersburg.

KUPFER (KUPFFER), Adolph-Theodor (Adol'f Yakovlevich), academician ordinarius for mineralogy from 27 Aug 1828; 11 Jan 1841 transferred to Chair of Physics (corresp member from 26 Dec 1826); Born: 6 Jan 1799 in Mitau; Died: 23 May 1865 in Petersburg.

KYOLER (KÖHLER), Heinrich-Karl-Ernst (Yegor Yegorovich), academician ordinarius for lit and Greco-Roman antiquities from 3 Sept 1817 (corresp member from 13 Apr 1803); Born: 25 Aug 1765 in Wechselburg, Saxony; Died: 22 Jan 1838 in Petersburg.

KYOL'REYTER (KOELREUTER), Joseph-Gottlieb, Dr of Med; junior associate for botany from 23 Dec 1755; left in mid 1761 (hon member from 1761); Born: 1733; Died: 30 Oct 1806 in Karlsruhe.

LAKSMAN (LAXMANN), Kirill (Erik) Gustavovich, prof of economics from 26 Feb 1770; acting prof of chemistry; left 18 Jan 1781 (hon member from 1781); Born: 24 July 1737 in Abo; Died: 5 Jan 1796 en route to Tobol'sk.

LAMANSKIY, Vladimir Ivanovich, academician ordinarius, Dept of Russian Language and Lit from 15 Jan 1900; Born: 26 June 1833 in Petersburg; Died: 15 Nov 1914 in Petersburg.

LANGSDORF (LANGSDORFF), Georg-Heinrich (Grigoriy Ivanovich), junior associate for botany from 18 July 1808; academician extraordinarius for zoology from 1 Apr 1812; left in 1831 (corresp member from 29 Jan 1803); Born: 18 Apr 1774 in Welstein, Rhine-Hessen; Died: 29 June 1852 in Freiburg, Breisgau.

* LAPPO-DANILEVSKIY, Aleksandr Sergeyevich, junior associate for history and Russian antiquities from 4 Dec 1899.

* LATYSHEV, Vasiliy Vasil'yevich, academician ordinarius for classical philology and archeology from 1 May 1893 (corresp member from 29 Dec 1890); Born: 29 July 1855 in vil Diyevo, Tver' Province; Died: 2 May 1921 in Petrograd.

LAVROVSKIY, Nikolay Alekseyevich, academician ordinarius, Dept of Russian Language and Lit from 3 Mar 1890 (corresp member from 29 Dec 1878); Born: 21 Nov 1825 in vil Vydropusk, Tver' Province; Died: 18 Sept 1899 in vil Kochetka, Khar'kov Province.

* LAZAREV, Pyotr Petrovich

LEKSEL' (LEXELL), Andreas-Johann, junior associate from 20 Mar 1769; prof of astronomy from 8 Apr 1771; Born: 24 Dec 1740 in Abo; Died: 30 Nov 1784 in Petersburg.

LEMAN (LEHMANN), Johann-Gottlieb, prof of chemistry from 1 Apr 1761; Born: 1700; Died: 11 Jan 1767 in Petersburg.

LENTS (LENZ), Heinrich-Friedrich-Emil (Emiliy Khristianovich), junior associate for physics from 7 May 1828; academician extraordinarius from 24 Mar 1830; academician ordinarius from 5 Sept 1834; Born: 12 Feb 1804 in Derpt; Died: 29 Jan 1865 in Rome.

LENTS (LENZ), Robert (Robert Khristianovich), junior associate for Sanskrit from 2 Oct 1835; Born: 23 Jan 1808 in Derpt; Died: 30 July 1836 in Petersburg.

LEPYOKHIN, Ivan Ivanovich, Dr of Med; junior associate from 23 May 1768; prof of natural history from 8 Apr 1771; Born: 10 Sept 1740 in Petersburg; Died: 6 Apr 1802 in Petersburg.

LERBERG (LEHRBERG), August-Christian, junior associate for history from 11 Mar 1807; academician extraordinarius from 7 Feb 1810; Born: 7 Aug 1770 in Derpt; Died: 23 July 1813 in Petersburg.

LERUA (LE ROY), Pierre-Louis, prof extraordinarius of modern history from 5 May 1735; left 10 Aug 1748 (hon member from 1748); Born: 15 Oct 1699 in Wesel, Duchy of Cleves; Died: 6 July 1774.

LEYTMAN (LEUTMANN), Johann-Georg, prof of mechanics and optics from 25 Apr 1726; Born: 30 Nov 1667 in Wittenberg, Saxony; Died: 5 Mar 1736 in Petersburg.

LIBERT (LIBERTUS), Johann-Christopher, prof of astronomy from 27 Feb 1736; dismissed 1 Sept 1737.

LOBANOV, Mikhail Yevstaf'yevich, academician ordinarius, Dept of Russian Language and Lit from 11 Jan 1845 (hon member from 21 Nov 1841; member, Russian Academy from 5 May 1828); Born: 8 Nov 1787 in Petersburg; Died: 5 June 1846 in Petersburg.

LOMONOSOV, Mikhail Vasil'yevich, junior associate for physics from 8 Jan 1742; prof of chemistry from 25 July 1745; Born: 8 Nov 1711 in vil of Denisovskaya, near Kholmogory; Died: 4 Apr 1765 in Petersburg.

LOTTER, Johann-Georg, prof of rhetoric and Greco-Roman antiquities from 1 Jan 1735; Born: 25 Mar 1702 in Augsburg; Died: 1 Apr 1737 in Petersburg.

LOVITS (LOWITZ), Georg-Moritz (Davyd Yegorovich), prof

of astronomy from Apr 1768; Born: 17 Feb 1722 in Fürth, Nuremberg; Died: 13 (24) Aug 1774 in Ilovlya on the Volga.

LOVITS (LOWITZ), Tobias-Johann (Toviy Yegorovich), junior associate for chemistry from 7 Oct 1790; academician ordinarius from 13 May 1793; Born: 25 Apr 1757 in Goettingen; Died: 26 Nov 1804 in Petersburg.

* LYAPUNOV, Aleksandr Mikhaylovich, academician ordinarius for applied mathematics from 6 Oct 1901 (corresp member from 29 Dec 1900); Born: 25 May 1857 in Yaroslavl'; Died: 3 Nov 1918 in Odessa.

MAKSIMOVICH, Karl Ivanovich, junior associate for botany from 8 Jan 1865; academician extraordinarius from 16 Feb 1868; academician ordinarius from 8 Jan 1871; Born: 11 Nov 1827 in Tula; Died: 4 Feb 1891 in Petersburg.

* MARKOV, Andrey Andreyevich, junior associate for pure mathematics from 13 Dec 1886.

* MARR, Nikolay Yakovlevich, junior associate for lit and history of Asian peoples from 7 Mar 1909.

MARTINI, Christian, prof of physics from 13 Jan 1725; prof of logic and metaphysics from 1726; dismissed from Academy 25 Jan 1729; Born: 1699 in Breslau; Died: after 1739.

MAYKOV, Leonid Nikolayevich, junior associate, Dept of Russian Language and Lit from 7 Jan 1889; academician extraordinarius from 14 Apr 1890; academician ordinarius from 2 Nov 1891 (corresp member from 29 Dec 1883); Born: 28 Mar 1839 in Petersburg; Died: 7 Apr 1900 in Petersburg.

MAYYER (MAYER), Friedrich-Christopher, junior associate from 1725; arrived with Bülffinger in the fall of 1725; prof extraordinarius of mathematics from 29 Jan 1726; Born: 9 Oct 1697 in Kirchheim unter Teck; Died: 24 Nov 1729 in Petersburg.

MERLING (MEURLING), Georg, junior associate and rector of the High School (Gymnasium) from 25 June 1736; left 25 Aug 1741; Born: in Sweden; Died: 8 Sept 1741 in Petersburg.

MERTENS, Karl Heinrich (Andrey Karlovich), junior associate for botany from 9 Sept 1829 and for zoology from 5 May 1830; Born: 17 May 1796 in Bremen; Died: 18 Sept 1830 in Petersburg.

MEYYER (MEYER), Karl-Anton (Karl Andreyevich), junior associate for botany from 27 Sept 1839; academician extraordinarius from 1844; academician ordinarius from 2 Aug 1845 (corresp member from 29 Dec 1833); Born: 20 Mar 1796 in Vitebsk; Died: 13 Feb 1855 in Petersburg.

MIDDENDORF (MIDDENDORFF), Alexander-Theodor (Aleksandr Fyodorovich), junior associate for zoology from 2 Aug 1845; academician extraordinarius from 2 Mar 1850; academician ordinarius from 1 Mar 1852; left 8 Mar 1865 (hon member from 5 Nov 1865); Born: 6 Aug 1815 in Liflyand; Died: 16 Jan 1894 on Gollenari Estate, Liflyand.

MIGIND (MYGIND), Franciscus, junior associate for chemistry from 1 Oct 1736; left 20 May 1737; Born: 1710 in Jutland; Died: 6 Apr 1789 in Vienna.

MIKHAYLOVSKIY-DANILEVSKIY, Aleksandr Ivanovich, academician ordinarius, Dept of Russian Language and Lit from 19 Oct 1841 (member, Russian Academy from 19 Sept 1831); Born: 26 Aug 1789 in Petersburg; Died: 9 Sept 1848 in Petersburg.

MILLER (MÜLLER), Gerhard-Friedrich (Fyodor Ivanovich), junior associate from 1725; prof of history from July 1730; conference secr from early 1728 to June 1730 and from Mar 1754 to 21 Feb 1765; Born: 18 Oct 1705 in Herford; Died: 11 Oct 1783 in Moscow.

MILLER, Vsevolod Fyodorovich, academician ordinarius, Dept of Russian Language and Lit from 5 Feb 1911 (corresp member from 29 Dec 1898); Born: 7 Apr 1848 in Moscow; Died: 5 Nov 1913 in Petersburg.

MODERAKH (MODERACH), Karl-Friedrich, junior associate from 13 June 1749; prof of history from Feb-Mar 1759; left 27 Aug 1761; rejoined 21 July 1763; Born: 11 Nov 1720 in Gross-Glogau; Died: 31 May 1772 in Petersburg.

MOISEYENKO (MOISEYENKOV), Fyodor Petrovich, junior associate for chemistry and mineralogy from 12 Oct 1779; Born: 11 Nov 1754 in Lebedyan', Khar'kov Province; Died: 1781 (or 1782) in Moscow.

MOTONIS, Nikolay Nikolayevich, junior associate from 15 Mar 1759 (hon member from 1767); Died: 20 Nov 1787 on Lipovyy Rog Estate, near Nezhin.

MULA (MOULA), Friedrich, junior associate for mathematics from 16 Jan 1736; left 10 July 1744; Born: in Neuchatel; Died: after 1749.

MURCHISON, Roderick-Imney, academician ordinarius for geology from 21 Sept 1845; Born: 19 Feb 1792 in Scotland; Died: 23 Oct 1871 in London.

* NASONOV, Nikolay Viktorovich

NASSE, Johann-Friedrich-Wilhelm, junior associate for technology from 27 Mar 1805; academician extraordinarius from 10 Oct 1810; left 2 Apr 1817; Born: 24 Dec 1780 in Bund, Westphalia.

NAUK (NAUCK), Avgust Karlovich, academician extraordinarius for classical philology from 6 June 1858; academician ordinarius from 2 June 1861; Born: 18 Sept 1822 in vil Auerstedt, near Jena; Died: 3 Aug 1892 in vil Terioki, near Petersburg.

NIKITENKO, Aleksandr Vasil'yevich, academician ordinarius, Dept of Russian Language and Lit from 20 Jan 1855 (corresp member from 29 Dec 1853); Born: Mar 1805 in vil Udarovka, Voronezh Province; Died: 20 July 1877 in Pavlovsk.

NIKITIN, Pyotr Vasil'yevich, junior associate for classical philology and archeology from 2 Apr 1888; academician extraordinarius from 22 Aug 1892; academician ordinarius from 18 Apr 1898; Born: 24 Jan 1849 in Ustyuzhna, Novgorod Province; Died: 5 May 1916 in Petrograd.

* NIKITSKIY, Aleksandr Vasil'yevich

* NIKOL'SKIY, Nikolay Konstantinovich

NOROV, Avraam Sergeyevich, academician ordinarius, Dept of Russian Language and Lit from 1 Nov 1851 (hon member from 21 Nov 1841); Born: 22 Oct 1795 in vil Klyuchi, Saratov Province; Died: 23 Jan 1869.

* OL'DENBURG, Sergey Fyodorovich, junior associate for

lit and history of Asian peoples from 5 Feb 1900.

OSTROGRADSKIY, Mikhail Vasil'yevich, junior associate for applied mathematics from 17 Dec 1828; academician extraordinarius from 11 Aug 1830; academician ordinarius from 21 Dec 1831; academician ordinarius for pure mathematics from 15 June 1855; Born: 12 Sept 1801 in vil Pashyonnaya, Poltava Province; Died: 20 Dec 1861 in Poltava.

OVSYANNIKOV, Filipp Vasil'yevich, junior associate for comparative anatomy and physiology from 2 Mar 1862; academician extraordinarius from 2 Aug 1863; academician ordinarius from 14 Aug 1864; Born: 14 June 1827 in Petersburg; Died: 29 May 1906 on Zapol'ye Estate, Petersburg Province.

OZERETSKOVSKIY, Nikolay Yakovlevich, Dr of Med; junior associate from 12 Oct 1779; academician ordinarius for natural history from 20 May 1782; Born: 1750 in Moscow Province; Died: 28 Feb 1827 in Petersburg.

PACHEKO, Rafael, junior associate for mechanics from 14 May 1764; Died: 31 Aug 1764.

* PALLADIN, Vladimir Ivanovich

* PAL'MOV, Ivan Savvich

PANAYEV, Vladimir Ivanovich, academician ordinarius, Dept of Russian Language and Lit from 19 Oct 1841 (member, Russian Academy from 6 May 1833); Born: 6 Nov 1792 in Turinsk; Died: 20 Nov 1859 in Khar'kov.

PANDER, Christian-Heinrich, junior associate for zoology from 7 June 1820; academician extraordinarius from 1823; academician ordianrius from 1 Feb 1826; left 30 June 1827; Born: 12 July 1794 in Riga; Died: 10 Sept 1865 in Petersburg.

PARROT, Georg-Friedrich (Yegor Ivanovich), academician ordinarius for applied mathematics from 26 Apr 1826; 24 Mar 1830 transferred to Chair of Physics (corresp member from 4 Dec 1811); left 29 Nov 1840 (hon member from 29 Dec 1840); Born: 5 July 1767 in Montbéliard; Died: 20 June 1852 in Helsingfors.

PAULAS, Peter-Simon (Pyotr Simonovich), prof of natural history from mid 1767; Born: 22 Sept 1741 in Berlin; Died: 8 Sept 1811 in Berlin.

* PAVLOV, Aleksey Petrovich

* PAVLOV, Ivan Petrovich

PAVSKIY, Gerasim Petrovich, academician ordinarius, Dept of Russian Language and Lit from 26 Feb 1858; Born: 3 Mar 1787 at Pava, Petersburg Province; Died: 7 Apr 1863 in Petersburg.

PEKARSKIY, Pyotr Petrovich, junior associate, Dept of Russian Language and Lit from 11 Jan 1863; academician extraordinarius from 1 May 1864; academician ordinarius from 4 Oct 1868; Born: 19 May 1827 near Ufa; Died: 12 July 1872 in Pavlovsk.

* PERETTS, Vladimir Nikolayevich

PEREVOSHCHIKOV, Dmitriy Matveyevich, junior associate for mathematics from 6 Mar 1852; academician extraordinarius from 20 Jan 1855 (corresp member from 29 Dec 1832); Born: 17 Apr 1788 in Saransk, Penza Province; Died: 3 Sept 1880 in Petersburg.

PETERS, Christian-August-Friedrich (Khristian Ivanovich), junior associate for theoretical astronomy from 5 Feb 1842; academician extraordinarius from 6 Mar 1847; left 4 Sept 1849 (corresp member from 4 Sept 1849); Born: 7 Sept 1806 in Hamburg; Died: 20 May 1880 in Kiel.

PETROV, Vasiliy Vladimirovich, junior associate for experimental physics from 11 Mar 1807; academician extraordinarius from 29 Nov 1809; academician ordinarius from 16 Aug 1815 (corresp member from 7 Feb 1802); Born: 8 July 1761 in Oboyan', Kursk Province; Died: 22 July 1834 in Petersburg.

PLATSMAN (PLATZMANN), Martin, junior associate for mathematics from 15 Jan 1784; Born: 1760 in Mol; Died: 25 Dec 1786 in Petersburg.

* PLESKE, Fyodor Dmitriyevich, junior associate for zoology from 24 Mar 1890.

PLETNYOV, Pyotr Aleksandrovich, academician ordinarius, Dept of Russian Language and Lit from 19 Oct 1841; Born: 10 Aug 1792 in Bezhitsa Uyezd; Died: 29 Dec 1865 in Paris.

POGODIN, Mikhail Petrovich, academician extraordinarius, Dept of Russian Language and Lit from 19 Oct 1841 (corresp member from 29 Dec 1826); member, Russian Academy from 14 Mar 1836; Born: 11 Nov 1800 in Moscow; Died: 8 Dec 1875 in Moscow.

POLYONOV, Vasiliy Alekseyevich, academician extraordinarius, Dept of Russian Language and Lit from 4 June 1842; hon member from 29 Dec 1833; Born: 1 Jan 1766 in vil Podberez'ya, near Valday; Died: 21 July 1851 near Petersburg.

POPOV, Nikita Ivanovich, junior associate from 19 Jan 1748; prof of astronomy from 1 Mar 1751; dismissed 5 Aug 1768; Born: 1720 in Yur'yev, Vladimir Province; Died: July 1782 in Voronezh.

PROTASOV, Aleksey Protas'yevich, junior associate from 22 Mar 1751; prof extraordinarius of anatomy from 1 Sept 1763; prof from 8 Apr 1771; Born: 1724; Died: 5 May 1796 in Petersburg.

PYPIN, Aleksandr Nikolayevich, academician ordinarius, Dept of Russian Language and Lit from 10 Jan 1898 (corresp member from 29 Dec 1891); Born: 25 Mar 1833 in Saratov Province; Died: 26 Nov 1904 in Petersburg.

* RADLOV, Vasiliy Vasil'yevich

REDOVSKIY, Ivan Ivanovich, junior associate for botany from 27 Mar 1805; Born: 1 Jan 1774 in Memel; Died: 8 Feb 1807 in Gizhiginsk.

RIKHMAN (RICHMANN), Georg-Wilhelm, junior associate from 15 Apr 1740; prof of physics from 20 Feb 1741; Born: 11 July 1711 in Pernov, Liflyand; Died: 26 July 1753 killed by lightning in Petersburg.

* ROSTOVTSEV, Mikhail Ivanovich, academician ordinarius for classical philology and archeology from 15 Apr 1917 (corresp member from 29 Dec 1908); Born: 28 Oct 1870 in Kiev; left in 1919.

ROZBERG, Mikhail Petrovich, junior associate, Dept of

Russian Language and Lit from 19 Oct 1841; academician extraordinarius from 3 Feb 1849; Born: 5 Aug 1804; Died: 1 Nov 1874 in Derpt.

ROZEN, Viktor Romanovich, junior associate for Oriental languages and lit from 16 Feb 1879; left 8 Mar 1882; rejoined as academician extraordinarius for history and lit of Asian peoples on 1 Dec 1890; academician ordinarius from 20 Jan 1901; Born: 21 Feb 1849 in Revel; Died: 10 Jan 1908 in Petersburg.

RUDOL'F (RUDOLPH), Johann-Friedrich (Ivan Yakovlevich), academician ordinarius for botany from 15 Feb 1804 (hon member from 28 Aug 1797); Born: 11 Jan 1754 in Jena; Died: 19 Aug 1809 in Petersburg.

RUMOVSKIY, Stepan Yakovlevich, junior associate from 18 Dec 1753; prof extraordinarius of astronomy from 1 Jan 1763; prof of astronomy from Jan 1767; vice-president from 3 Nov 1800 (hon member from 1767); left in June 1803; Born: 29 Oct 1734 in vil of Staryy Pogost, Vladimir Province; Died: 6 July 1812 in Petersburg.

RUPREKHT (RUPRECHT), Frants Ivanovich, junior associate for botany from 5 Feb 1848; academician extraordinarius from 5 Nov 1853; academician ordinarius from 11 Jan 1857; Born: 1 Nov 1814 in Freiburg, Breisgau; Died: 23 July 1870 in Petersburg.

* RYKACHYOV, Mikhail Aleksandrovich

SAL'KHOV (SALCHOW), Ulrich-Christopher, prof of chemistry from 23 Dec 1755; left prior to 29 May 1760; Born: 9 Feb 1722 in Kaznewitz, Rügen Island; Died: 20 Apr 1787 in Mehldorf.

SAVVICH, Aleksey Nikolayevich, academician extraordinarius for astronomy from 1 June 1862; academician ordinarius from 10 May 1868 (corresp member from 29 Dec 1852); Born: 18 Mar 1811 in Belovodsk, Khar'kov Province; Died: 15 Aug 1883 in Tula Province.

SEVAST'YANOV, Aleksandr Fyodorovich, junior associate for natural history from 4 Nov 1799; academician extraordinarius from 14 Aug 1803; academician ordinarius from 19 Dec 1810; Born: 1771; Died: 5 Dec 1824.

SEVERGIN, Vasiliy Mikhaylovich, junior associate for

mineralogy from 25 June 1789; academician ordinarius from 6 May 1793; Born: 8 Apr 1765 in Petersburg; Died: 17 Nov 1826 in Petersburg.

* SHAKHMATOV, Aleksey Aleksandrovich, junior associate, Dept of Russian Language and Lit from 12 Nov 1894.

SHARMUA (CHARMOY), Francois-Bernard (Frants Frantsevich), junior associate for Oriental languages and antiquities from 16 May 1832; left 1 Sept 1835 (corresp member from 29 Dec 1829 to 1832 and from 1 Sept 1835 to 1868); Born: 1793; Died: 9 Dec 1868 in Aoust, Department of Drome, France.

SHERER (SCHERER), Alexander-Nikolas (Aleksandr Ivanovich), junior associate for chemistry from 27 Mar 1805; academician extraordinarius from 11 Mar 1807; academician ordinarius from 16 Aug 1815 (corresp member from 1796); Born: 30 Dec 1771 in Petersburg; Died: 17 Oct 1824 in Petersburg.

SHEVYRYOV, Stepan Petrovich, junior associate, Dept of Russian Language and Lit from 19 Oct 1841; academician extraordinarius from 6 Feb 1847; academician ordinarius from 6 Nov 1852; Born: 18 Oct 1806 in Saratov; Died: 8 May 1864 in Paris.

SHIFNER (SCHIEFNER), Franz-Anton (Anton Antonovich), junior associate for Tibetan language from 5 June 1852; academician extraordinarius from 3 June 1854; Born: 6 July 1817 in Revel; Died: 4 Nov 1879 in Petersburg.

SHIRINSKIY-SHIKHMATOV, Platon Aleksandrovich, academician ordinarius, Dept of Russian Language and Lit from 19 Oct 1841 (hon member from 29 Dec 1837; member, Russian Academy from 5 May 1828); Born: 1790 in vil Dernovo, Smolensk Province; Died: 5 May 1853 in Petersburg.

SHLEGEL'MIL'KH (SCHLEGELMILCH), Aleksandr Karlovich, junior associate for mineralogy from 14 Dec 1808; academician extraordinarius from 1 Apr 1812; left 14 Aug 1820; Born: 1777; Died: after 1830.

SHLYOTSER (SCHLÖZER), August-Ludwig, junior associate from 1 June 1762; prof of history from 3 Jan 1765;

left 30 Jan 1769 (hon member from 1769); Born: 5 July 1735 in Jagdstadt, Hohenlohe; Died: 9 Sept 1809 in Goettingen.

SHMIDT (SCHMIDT), Fyodor Bogdanovich, junior associate for paleontology from 7 Jan 1872; academician extraordinarius from 3 May 1874; academician ordinarius from 13 Apr 1885; Born: 15 Jan 1832 on Kaisma Estate, Liflyand Province; Died: 8 Nov 1908 in Petersburg.

SHMIDT (SCHMIDT), Isaac-Jacob (Yakov Ivanovich), junior associate for lit and antiquities of the Orient from 28 Jan 1829; academician extraordinarius from 9 Mar 1831; academician ordinarius from 14 June 1833 (corresp member from 25 Oct 1826); Born: 14 Oct 1779 in Amsterdam; Died: 27 Aug 1847 in Petersburg.

SHMIDT (SCHMIDT), Jacob-Friedrich, junior associate, Geographical Dept, Academy of Sci from 11 July 1757; Died: after 1780 in Petersburg.

SHRENK (SCHRENCK), Leopold (Leopol'd Ivanovich), junior associate for zoology from 2 Mar 1862; academician extraordinarius from 2 Aug 1863; academician ordinarius from 4 June 1865; Born: 24 Apr 1826 on Khoten' Estate, Khar'kov Province; Died: 8 Jan 1894 in Petersburg.

SHTELIN (STÄHLIN), fon Yakov Yakovlevich, junior associate from 1 Feb 1735; arrived 24 June 1735; prof of rhetoric and poetry from 2 Sept 1737; conference secr from 7 Mar 1765 to 22 Feb 1769; Born: 25 Apr 1709 in Memmingen, Swabia; Died: 25 June 1785 in Petersburg.

SHTELLER (STELLER), Georg-Wilhelm, junior associate for natural history from 7 Feb 1737; Born: 10 Mar 1709 in Winzenheim, Franconia; Died: 12 Nov 1746 in Tyumen'.

SHTORKH (STORCH), Heinrich (Andrey Karlovich), academician ordinarius for political economy and statistics from 1 Feb 1804 (corresp member from 7 Apr 1796); Born: 1 Mar 1766 in Riga; Died: 1 Nov 1835 in Petersburg.

SHTRAUKH (STRAUCH), Aleksandr Aleksandrovich, junior associate for zoology from 7 Apr 1867; academician extraordinarius from 1 May 1870; academician ordinarius from 2 Nov 1879; Born: 1 Mar 1832 in Petersburg; Died: 14 Aug 1893 in Wiesbaden.

SHTRUBER-PIRMONT (STRUBER DE PIERMONT), Friedrich-Heinrich, prof of law from 22 Sept 1738; left 23 Feb 1741; rejoined 1 July 1746; secr from 2 July 1746 to 1 Mar 1749; left again 7 July 1757 (1771-1775 councillor to Collegium of Foreign Affairs); Born: 1704 in Piermont; Died: prior to 1790.

SHUBERT (SCHUBERT), Friedrich-Theodor (Fyodor Ivanovich), junior associate from 18 Sept 1786; academician ordinarius for mathematics from 18 June 1789; academician ordinarius for astronomy from 1803; Born: 19 (30) Oct 1758 in Helmstedt; Died: 10 Oct 1825 in Petersburg.

SHYOGREN (SJÖGREN), Johann-Andreas (Andrey Mikhaylovich), junior associate for history from 30 Sept 1829; academician extraordinarius for history and Russian antiquities from 9 Mar 1831; academician ordinarius for the ethnography and languages of Finnish and Caucasian tribes from 5 Oct 1844 (corresp member from 29 Dec 1827); Born: 26 Apr 1794 in vil Stikkalya, Neuland Province; Died: 6 Jan 1855 in Petersburg.

SIGEZBEK (SIEGESBECK), Johann-Georg, prof of botany and natural history from 5 Apr 1742; left 1 May 1747; Born: circa 1689 in Wittenberg; Died: Jan 1755 in Seehausen.

SMELOVSKIY, Timofey Andreyevich, junior associate for botany from 19 May 1802; academician extraordinarius from 14 Aug 1803; academician ordinarius from 16 Aug 1815; Born: 1769 in the Ukraine; Died: 21 Oct 1815 in Petersburg.

* SMIRNOV, Yakov Ivanovich

* SOBOLEVSKIY, Aleksey Ivanovich

SOFRONOV, Mikhail, junior associate for mathematics from 23 Dec 1753; Born: 1729 in Ustyuzhen', Novgorod Province; Died: 1761.

SOKOLOV, Nikita Petrovich, Dr of Med; junior associate for chemistry from 10 Mar 1783; academician ordinarius from 27 Sept 1787; left 30 Sept 1792; (hon member from

1792; member, Russian Academy from 1784); Born: 1748; Died: 7 Apr 1795 in Moscow.

SOLOV'YOV, Sergey Mikhaylovich, academician ordinarius, Dept of Russian Language and Lit from 3 Mar 1872 (corresp member from 29 Dec 1864); Born: 5 May 1820 in Moscow; Died: 4 Oct 1879 in Moscow.

SOMOV, Osip Ivanovich, academician ordinarius for pure mathematics from 2 Mar 1862 (corresp member from 29 Dec 1852); Born: 1 June 1815 in Klin Uyezd, Moscow Province; Died: 26 Apr 1876 in Petersburg.

SONIN, Nikolay Yakovlevich, academician ordinarius for pure mathematics from 1 Mar 1893 (corresp member from 29 Dec 1891); Born: 10 Feb 1849 in Tula; Died: 14 Feb 1915 in Petrograd.

SREZNEVSKIY, Izmail Ivanovich, junior associate, Dept of Russian Language and Lit from 3 Feb 1849; academician extraordinarius from 1 Nov 1851; academician ordinarius from 4 Nov 1854; Born: 1 June 1812 in Yaroslavl'; Died: 9 Feb 1890 in Petersburg.

STEFANI (STEPHANI), Ludolph (Ludol'f Eduardovich), academician ordinarius for Greco-Roman antiquities from 7 Sept 1850; Born: 29 Mar 1816 in vil Beich, near Leipzig; Died: 30 May 1887 in Pavlovsk.

* STEKLOV, Vladimir Andreyevich, junior associate for applied mathematics from 6 Nov 1910.

STRITTER, Johann-Gotthilf, junior associate for history from 10 Oct 1779 (hon member from 1787); Born: 10 Oct 1740 in Idstein, Nassau; Died: 19 Feb 1801 in Moscow.

STROYEV, Pavel Mikhaylovich, junior associate, Dept of Russian Language and Lit from 19 Oct 1841; academician extraordinarius from 6 Feb 1847; academician ordinarius from 3 Feb 1849 (corresp member from 29 Dec 1826); Born: 27 June 1796; Died: 5 Jan 1876 in Moscow.

STRUVE, Friedrich-Georg-Wilhelm (Vasiliy Yakovlevich), academician extraordinarius for astronomy from 18 Jan 1832; left 21 Dec 1861 (corresp member from 29 Jan 1822; hon member from 29 Dec 1826 and again from 1861); Born: 15 Apr 1793 in Altona; Died: 11 Nov 1864 in Petersburg.

STRUVE, Otto-Wilhelm (Otton Vasil'yevich), junior associate for astronomy from 4 Dec 1852; academician extraordinarius from 10 Feb 1856; academician ordinarius from 19 Apr 1861; left 16 Dec 1889; Born: 25 Apr 1819 in Derpt; Died: 1 Apr 1905 in Karlsruhe.

* STRUVE, Pyotr Berngardovich

SUKHOMLINOV, Mikhail Ivanovich, academician extraordinarius, Dept of Russian Language and Lit from 3 Nov 1872; academician ordinarius from 6 Feb 1876 (corresp member from 29 Dec 1855); Born: 3 Mar 1828 in Khar'kov; Died: 8 July 1901 in Petersburg.

TARKHANOV, Pavel Vasil'yevich, junior associate for astronomy from 9 Oct 1822; academician extraordinarius from 26 Apr 1826; Born: 29 Oct 1787 in Uglich, Yaroslavl' Province; Died: 16 Mar 1839 in Petersburg.

TAUBERT, Johann-Kaspar (Ivan Ivanovich), junior associate for history from 29 May 1738; asst librarian; Born: 31 Aug 1717 in Petersburg; Died: 9 May 1771 in Petersburg.

TEPLOV, Grigoriy Nikolayevich, junior associate for botany from 3 Jan 1742; left 7 Mar 1743; rejoined Apr 1743; assessor, Academic Chancellery from 1 July 1746; member, Academy Assembly from 28 July 1747 (hon member from 1747); Born: 1711 (?) in Pskov; Died: 13 Mar 1779 in Petersburg.

TIKHONRAVOV, Nikolay Savvich, academician ordinarius, Dept of Russian Language and Lit from 3 Mar 1890 (corresp member from 29 Dec 1863); Born: 3 Oct 1832 in vil Shemetovo, Kaluga Province; Died: 27 Nov 1893 in Moscow.

TILEZIUS FON TILENAU (TILESIUS VON TILENAU), Wilhelm-Gottlieb, junior associate for natural history from 15 Oct 1806; academician extraordinarius from 12 Apr 1809; left 10 Sept 1817 (hon member from 17 Sept 1817); Born: 17 July 1769 in Mülhausen, Saxony; Died: 17 Mar 1857 in Mülhausen.

TREDIAKOVSKIY, Vasiliy Kirillovich, prof of rhetoric from 25 July 1745; dismissed 30 Mar 1759; Born: 22 Feb in Astrakhan'; Died: 6 Aug 1769 in Petersburg.

TRINIUS, Karl-Bernhard (Karl Antonovich), academician extraordinarius for botany from 30 Apr 1823 (corresp member from 30 May 1810); Born: 7 Mar 1778 in Eisleben, Saxony; Died: 29 Feb 1844 in Petersburg.

TRUSKOTT (TROSCOTT, TRESCOTT), Ivan Fomich, junior associate for geography from 18 June 1742; Born: 1 Dec 1719 in Petersburg; Died: 18 May 1786 in Petersburg.

TSEYGER (ZEIGER), Johann-Ernst, prof of mechanics from Mar 1756; left July 1764 (hon member from 1764); Born: 1720 in Weissenfels; Died: 7 Jan 1784 in Dresden.

* USPENSKIY, Fyodor Ivanovich

USTRYALOV, Nikolay Gerasimovich, junior associate for history and Russian antiquities from 13 Jan 1837; academician extraordinarius from 4 June 1842; academician ordinarius from 5 Oct 1844; Born: 4 May 1805 in Maloarkhangel'sk Uyezd, Oryol Province; Died: 8 June 1870 in Tsarskoye Selo.

VAL'DEN, Pavel Ivanovich, academician ordinarius for technology and chemistry applied to sciences and crafts from 1 May 1910; Born: 14 July 1863 on Rosenbeck Estate, near Wenden, Liflyand Province; Died: 22 Jan 1957 abroad.

VASILEVSKIY, Vasiliy Grigor'yevich, academician ordinarius for Russian and Byzantine history from 3 Feb 1890 (corresp member from 29 Dec 1876); Born: 21 Jan 1838 in vil Il'inskoye, Yaroslavl' Province; Died: 13 May 1899 in Florence.

VASIL'YEV, Vasiliy Pavlovich, academician ordinarius for lit and history of Asian peoples from 11 Jan 1886 (corresp member from 29 Dec 1866); Born: 20 Feb 1818 in Nizhniy Novgorod; Died: 27 Apr 1900 in Petersburg.

VEL'YAMINOV-ZERNOV, Vladimir Vladimirovich, junior associate for lit and history of Asian peoples from 6 June 1858; academician extraordinarius from 1 Dec 1861; left 15 Feb 1878 (hon member from 29 Dec 1890); Born: 31 Oct 1830 in Petersburg; Died: 17 Jan 1904 in Kiev.

* VERNADSKIY, Vladimir Ivanovich, junior associate for mineralogy from 4 Mar 1906.

VESELOVSKIY, Aleksandr Nikolayevich, junior associate, Dept of Russian Language and Lit from 2 Dec 1877; academician extraordinarius from 2 May 1880; academician ordinarius from 4 Dec 1881 (corresp member from 29 Dec 1876); Born: 4 Feb 1838 in Moscow; Died: 10 Oct 1906 in Petersburg.

VESELOVSKIY, Konstantin Stepanovich, junior associate for statistics and political economy from 1 May 1852; academician extraordinarius from 1 Sept 1855; academician ordinarius from 5 June 1859; Born: 20 May 1819 in Yekaterinoslav Province; Died: 3 Nov 1901 in Petersburg.

VEYTBREKHT (WEITBRECHT), Josia, junior associate from 1725; arrived with Du Vernoi in Dec 1725; prof of physiology from 1 Jan 1731; Born: 20 Oct 1702 in Schorndorf, Württemberg; Died: 8 Feb 1747 in Petersburg.

VIDEMAN (WIEDEMANN), Ferdinand-Johann (Ferdinand Ivanovich), academician extraordinarius for the languages and ethnography of Finnish tribes from 10 Sept 1857; academician ordinarius from 2 Oct 1859; Born: 18 Mar 1805 in Khapsala; Died: 17 Dec 1887 in Petersburg.

VIL'D (WILD), Genrikh Ivanovich, academician extraordinarius for physics and meteorology from 10 May 1868; academician ordinarius from 1 May 1870; left 16 Sept 1895 (hon member from 29 Dec 1895); Born: 17 Dec 1833 in Zurich; Died: 5 Sept 1902 in Zurich.

VIL'DE (WILDE), Johann-Christian, junior associate from 27 Feb 1736; prof extraordinarius of anatomy from 2 May 1738; left 5 July 1744; Born: in Züllichau, Prussia.

* VINOGRADOV, Pavel Gavrilovich

VINZGEYM (WINZHEIM), Christian-Nikolas, junior associate from 1 May 1731; prof extraordinarius of astronomy from 1 Jan 1735; conference secr from 1742 to 1746 and from 1749 to 1751; Born: in Anklam, Prussia; Died: 4 Mar 1751 in Petersburg.

VISHNEVSKIY, Vikentiy Karlovich, junior associate for astronomy from 15 Feb 1804; academician extraordinarius from 11 Mar 1807; academician ordinarius from 15

Feb 1815; Born: 1781 in Poland; Died: 1 June 1855 in Petersburg.

VISKOVATOV, Vasiliy Ivanovich, junior associate for mathematics from 14 Aug 1803; academician extraordinarius from 11 Mar 1807 (corresp member from 1 July 1799); Born: 26 Dec 1779 in Petersburg; Died: 8 Oct 1812 in Petersburg.

VOL'F (WOLFF), Kaspar-Friedrich, prof of anatomy and physiology from mid 1767; Born: 1733 in Berlin; Died: 22 Feb 1794 in Petersburg.

VOLKOV, Aleksey Gavrilovich, junior associate for chemistry from 14 Dec 1803; left 18 Oct 1809; Born: 1780; Died: after Apr 1825.

VORONIN, Mikhail Stepanovich, academician ordinarius for botany from 10 Jan 1898 (corresp member from 29 Dec 1894); Born: 21 June 1838 in Petersburg; Died: 20 Feb 1903 in Petersburg.

VOSTOKOV, Aleksandr Khristoforovich, academician ordinarius, Dept of Russian Language and Lit from 19 Oct 1841 (corresp member from 26 Dec 1826; member, Russian Academy from 12 June 1820); Born: 16 Mar 1781 in Arensburg, Esel Island; Died: 8 Feb 1864 in Petersburg.

VOVIL'YE (VAUVILLIERS), Jean-Francois, academician ordinarius for history and antiquities from 7 June 1798; Born: 24 Sept 1737 in Noyers; Died: 23 July 1801 in Petersburg.

VYAZEMSKIY, Pyotr Andreyevich, academician ordinarius, Dept of Russian Language and Lit from 19 Oct 1841 (member, Russian Academy from 2 Oct 1839); Born: 12 July 1792 in Moscow; Died: 22 Nov 1878 in Baden-Baden.

* YAGICH (JAGIC), Vatroslav (Ignatiy Vikent'yevich)

YAKOBI (JACOBI), Moritz-Hermann (Boris Semyonovich), junior associate, Chair of Practical Mechanics and Theory of Machines from 29 Nov 1839; academician extraordinarius from 7 May 1842; academician ordinarius for technology and applied chemistry from 6 Mar 1847, and for physics from 21 Sept 1865 (corresp member from 29 Dec 1838); Born: 21 Sept 1801 in Potsdam; Died: 27 Feb 1874 in Petersburg.

YANZHUL, Ivan Ivanovich, academician ordinarius for political economy and financial science from 4 Mar 1895 (corresp member from 29 Dec 1893); Born: 2 June 1846 in Pyatigory, Kiev Province; Died: 31 Oct 1914 in Wiesbaden.

YARTSOV, Yanuariy Osipovich, junior associate for Oriental languages from 10 June 1818; left in July 1819; Born: 1792; Died: 3 Dec 1861 in Petersburg.

YAZYKOV, Dmitriy Ivanovich, academician ordinarius, Dept of Russian Language and Lit from 19 Oct 1841 (hon member from 29 Dec 1830; member, Russian Academy from 7 Jan 1833); Born: 14 Oct 1773 in Moscow; Died: 13 Nov 1845 in Petersburg.

YEREMEYEV, Pavel Vladimirovich, academician extraordinarius for mineralogy from 3 Sept 1894 (corresp member from 29 Dec 1875); Born: 1830 in Tobol'sk Province; Died: 6 Jan 1899 in Petersburg.

YUNKER (JUNCKER), Gottlob-Friedrich-Wilhelm, junior associate from late 1731; prof of politics and ethics from 23 Apr 1734; prof of rhetoric from 25 Nov 1734; left 25 Aug 1737 (hon member from 1737); Born: 1702 in Schleusingen (according to other sources - 1 July 1705 in Altenburg, Saxony); Died: 10 Nov 1746 in Petersburg.

ZAGORSKIY, Pyotr Andreyevich, junior associate for anatomy and physiology from 27 Mar 1805; academician extraordinarius from 11 Mar 1807; academician ordinarius from 18 Nov 1807 (hon member, Dept of Russian Language and Lit from 21 Nov 1841); Born: 9 Aug 1764 in Podgornitsa, near Novgorod-Severskiy, Chernigov Province; Died: 20 Mar 1846 in Petersburg.

ZAKHAROV, Yakov Dmitriyevich, junior associate for chemistry from 3 May 1790; prof extraordinarius from 29 Jan 1795; academician ordinarius from 31 Jan 1798; Born: 3 Oct 1765 in Petersburg; Died: 2 Oct 1836 in Petersburg.

ZALEMAN, Karl Germanovich, junior associate for lit and history of Asian peoples from 16 Aug 1886; academician extraordinarius from 4 Nov 1889; academician ordinarius from 4 Feb 1895; Born: 28 Dec 1849 in Revel; Died: 30 Nov 1916 in Petrograd.

* ZALENSKIY, Vladimir Vladimirovich, academician ordinarius for zoology from 18 Jan 1897 (corresp member from 29 Dec 1893); Born: 26 Jan 1847 in vil Shakhvorostovka, Poltava Province; Died: 8 Oct 1918 in Sebastopol.

ZHDANOV, Ivan Nikolayevich, academician ordinarius, Dept of Russian Language and Lit from 4 Dec 1899 (corresp member from 29 Dec 1893); Born: 22 June 1846 in Shenkorsk; Died: 11 July 1901 in Yalta.

ZHELEZNOV, Nikolay Ivanovich, junior associate for plant physiology as applied to agriculture from 14 May 1853; academician extraordinarius from 7 June 1857; Born: 22 Oct 1816 in Petersburg; Died: 15 Jan 1877 in Petersburg.

ZHUKOVSKIY, Vasiliy Andreyevich, academician ordinarius, Dept of Russian Language and Lit from 19 Oct 1841 (hon member from 29 Dec 1827; member, Russian Academy from 5 Oct 1818); Born: 29 Jan 1783 in vil Mishenskoye, near Belev, Tula Province; Died: 12 Apr 1852 in Baden-Baden.

ZININ, Nikolay Nikolayevich, junior associate for chemistry from 2 June 1855; academician extraordinarius from 2 May 1858; academician ordinarius for technology and chemistry applied to arts and crafts from 5 Nov 1865; Born: 13 Aug 1812 in Shusha; Died: 6 Feb 1880 in Petersburg.

ZOLOTARYOV, Yegor Ivanovich, junior associate for mathematics from 3 Dec 1876; academician extraordinarius from 2 May 1878; Born: 29 Mar 1847 in Petersburg; Died: 7 July 1878 in Petersburg.

ZUYEV, Vasiliy Fyodorovich, junior associate from 12 Oct 1779; prof of natural history from 27 Sept 1787; Born: 1 Jan 1754 in Petersburg; Died: 7 Jan 1794 in Petersburg.

Foreign Members of the USSR Academy of Sciences

year of election

Akabori, Siro, biochemist, Japan	1966
Alven, Hannes, physicist, Sweden	1958
Amaldi, Edoardo, physicist, Italy	1958
Bac, Zenon Marcel, biologist, Belgium	1958
Bandinelli, Ranucco Bianci, art historian, Italy	1958
Bernal, John Desmond, physicist, England	1958
Blackett, Patrick Maynard Stewart, nuclear physicist, England	1966
Blašković, Dionis, virologist, Czechoslovakia	1966
Born, Max, physicist, GFR	1934
Bronk, Detlef Wulf, biologist, USA	1958
Chan Dai Ngia, mechanic engineer, North Vietnam	1966
Ciceyka, Sherban, physicist, Rumania	1966
Courant, Richard, mathematician, USA	1966
De Broil, Louis, physicist, France	1958
Dirac, Paul, physicist, England	1931
Draiche, Jean, geographer, France	1966
Erdei-Gruz, Tibor, chemist, Hungary	1966
Florey, Howard Walter, physiologist, England	1966
Groszowski, Janesz, physicist, Poland	1966
Hartke, Werner, historian, GDR	1966
Herz, Gustav, physicist, GDR	1958
Hinshelwood, Cyril Norman, chemist, England	1958
Jablonski, Henryk, historian, Poland	1966
Kaya, Seiji, physicist, Japan	1958
Ko Mo-jo, historian, Chinese People's Republic	1958
Kotarbinski, Tadeusz, philosopher, Poland	1958
Krystanov, Lyubomir, geophysicist, Bulgaria	1966
Kuratowski, Kasimir, mathematician, Poland	1966
Lee Sing Gee, chemist, North Korea	1966
Leraix, Jean, mathematician, France	1966
Li Sze-kwang, geologist, Chinese People's Republic	1958
Mahalanobis Prasanta Chandra, statiscian, India	1958
Mark, Herman Francis, chemist, USA	1966
Murgulescu, Ilie, chemist, Rumania	1966
Nadzhakov, Georgiy, physicist, Bulgaria	1958
Natta, Julio, chemist, Italy	1966
Neel, Louis Boyd, physicist, France	1958
Nenicescu, Kostin, chemist, Rumania	1966
Ochoa, Severo, biochemist, USA	1966

Oort, Jan Hendrik, astronomer, Holland	1966
Parhon, Constantin, endocrinologist, Rumania	1947
Pauling, Linus, chemist, USA	1958
Pavlov, Todor, literary historian, Bulgaria	1947
Pek Nam Un, historian, North Korea	1958
Penfield, Wilder Graves, physician, Canada	1958
Powell, Cecil Frank, physicist, England	1958
Prelog, Wladimir, chemist, Switzerland	1966
Raman Chandrasekhara Venkata, physicist, India	1947
Rinecker, Günter, chemist, GDR	1966
Robinson, Robert, chemist, England	1966
Rusnyak, Istvan, physician, Hungary	1958
Ruzhichka, Leopold, chemist, Switzerland	1958
Saint-Dierdi, Albert, biochemist, USA	1947
Savić, Pavle, Yugoslavia	1958
Shirendev Bazaryn, historian, Mongolia	1966
Shorm, František, organic chemist, Czechoslovakia	1958
Sigban, Manne, physicist, Sweden	1958
Stanković, Sinisha, biologist, Yugoslavia	1966
Steenbeck, Max, physicist, GDR	1966
Stenshe, Eric, paleontologist, Sweden	1929
Swedberg, Theodor, chemist, Sweden	1966
Taylor, Jeffrey Ingram, mechanical engineer, England	1966
Thyssen, Peter-Adolf, chemist, GDR	1966
Timoshenko, Stepan Prokof'yevich, mechanical engineer, USA	1928
Turki Akhmed Riad, chemist, UAR	1958
Watson, David M.S., paleontologist, England	1932
Yukawa, Hideki, physicist, Japan	1966

SUPPLEMENTS

International Organizations

The USSR Academy of Sciences is a member of the following international scientific organizations:

International Council of Scientific Unions

Scientific Committee for Oceanographic Research, International Council of Scientific Unions

Scientific Committee for Antarctic Research, International Council of Scientific Unions

Scientific Committee for Conducting the International Year of the Quiet Sun, International Council of Scientific Unions

Scientific Committee for Space Research, International Council of Scientific Unions

Special Committee for the International Biological Program, International Council of Scientific Unions

International Geophysical Committee

International Reviewing Bureau

International Documentation Federation

International Astronomical Union

International Crystallographic Union

International Geodetical and Geophysical Union

International Geodetical Association

International Association of Seismology and Physics of the Earth's Interior

International Association of Meteorology and Atmospheric Physics

International Association of Geomagnetism and Aeronomy

International Oceanographic Association

International Association of Scientific Hydrology

International Association of Volcanology

International Mathematics Union

International Union of Theoretical and Applied Physics

International Astronautical Federation

International Academy of Astronautics

International Cybernetical Association

International Association for Modelling Devices

International Committee for Thermodynamics and Electrochemical Kinetics

International Committee for High-Speed Photography

International Federation of Information Processing Societies

Permanent Council for Arranging International Congresses on Corrosion

International Union of Theoretical and Applied Chemistry

International Geographical Union

International Society of Soil Science

International Hydrogeologists' Association

International Commission for Geological Mapping of the World

International Commission for Geological Mapping of Europe

International Association for Deep Zones of the Earth's Crust

International Stratigraphic Commission

International Association for the Study of Loams

Commission for the Study of Meteorites

International Organization for Sedimentology

Carpatho-Balkan Geological Association

International Geological Union

International Biochemical Union

International Society for Cell Biology

International Union of Physiological Sciences

International Federation of Electroencephalography and Clinical Neurophysiology Societies

International Institute of Embryology

International Union of Forest Research Organizations

International Union of Biological Sciences

Permanent Committee for Convening Entomological Congresses

European Nematodological Society

International Society of Horticultural Sciences

International Limnological Society

International Association for Plant Taxonomy

International Association of Botanical Gardens (Part of the Plant Taxonomy Association)

International Union of Theoretical and Applied Biophysics

International Brain Research Organization

International Society for Clinical Electroretinography

International Cell Research Organization

International Union of Theoretical and Applied Mechanics

International Scientific Radio Union

International Federation of Automatic Control

International Steam Research Commission

International Institute of Welding

International Council for Convening Oil Congresses

Working Group for International Critical Tables, International Council of Scientific Unions

International Biometric Society

International Union of Orientalists

International Union of Anthropologists

International Committee of Historical Sciences

International Association for Iranian Inscriptions

International Center for Ancient Fabrics

International Sociological Institute

International Union of Prehistorical and Protohistorical Sciences

International Association of Byzantologists

International Union of History and Philosophy of Science

International Academy of the History of Science

International Association of Criminal Law

International Association of Economic Sciences

International Committee of State Finances

International Association of Legal Sciences

European Center for the Coordination of Research and Documentation in the Social Sciences

International Statistical Institute

International Law Association

International Institute of Space Law

International Federation of Philosophical Societies

International Sociological Association

International Association of Political Sciences

International Committee for the Documentation of Social Sciences

Institute of International and Comparative Agrarian Law

International Academy of the Philosophy of Sciences

International German Language and Literature Association

International Committee of Slavists

International Association for the Study of Southeast Europe

International Institute of Administrative Sciences

International Association for Population Problems

International Association for Economic History

International Folklore Society*

---

* On 1 Jan 1968 there were eight Soviet scientists or scholars functioning as presidents of international unions and 43 as vice-presidents. Another 58 were members of executive committees of international scientific organizations.

National Committees and Associations of Soviet Scientists

USSR National Oil Committee (International Council for Convening World Oil Congresses)

Interdepartmental Geophysical Committee (International Geodetic and Geophysical Union)

Soviet Coordination Committee for the Properties of Steam (International Commission for the Study of Steam)

Soviet National Committee of the International Scientific Radio Union (International Scientific Radio Union)

Interdepartmental Commission for the Study of the Antarctic (Scientific Committee for Antarctic Studies)

National Committee of Soviet Physicists (International Union of Theoretical and Applied Physics)

National Committee of Soviet Crystallographers (International Crystallographic Union)

National Committee of Soviet Mathematicians (International Mathematics Union)

Astronomical Council (International Astronomical Union)

National Committee of Geologists of the Soviet Union (International Geological Union)

National Committee of Soviet Geographers (International Geographical Union)

National Committee of Soviet Biologists (International Union of Biological Sciences)

National Committee of Soviet Biochemists (International Biochemical Union)

USSR National Committee of Automatic Control (International Federation of Automatic Control)

USSR National Welding Committee (International Institute of Welding)

USSR National Committee for Theoretical and Applied Mechanics (International Union of Theoretical and Applied Mechanics)

National Committee of Historians of the Soviet Union (International Committee of Historical Sciences)

Soviet Committee of Slavists (International Committee of Slavists)

Association of Soviet Economic Scientific Establishments (International Association of Economic Sciences)

Soviet Pugwash Committee

Scientific Council for the "Sun-Earth" Problem

Soviet National Committee of the International Brain Research Organization

National Committee of Soviet Chemists

Soviet Association of International Law (International Law Association)

Soviet Political Sciences Association (International Association of Political Sciences)

Soviet Sociological Association (International Sociological Association)

National Committee for Collating and Processing Numerical Data in Science and Engineering

Soviet National Association of Historians of Natural Sciences and Engineering (International Association for the History and Philosophy of Science)

USSR National Committee for Oceanography

Soviet National Committee for Conducting the International Biological Program

Soviet National Committee of the Pacific Scientific Association

USSR National Cartographers Council*

---

* National committees (or associations) cover those specialists appointed by the USSR Academy of Sciences to function in international scientific organizations.

Gold Medals and Prizes Awarded by the USSR Academy of Sciences*

V.V. Dokuchayev Gold Medal, awarded to Soviet and foreign scientists for work and discoveries in the field of soil science.

A.P. Karpinskiy Gold Medal, awarded to Soviet and foreign scientists for aggregate work in geology, paleontology, petrography and minerals.

S.P. Korolyov Gold Medal, awarded to Soviet scientists for work on rocketry and space engineering.

I.V. Kurchatov Gold Medal, with cash award, conferred on Soviet scientists for work on nuclear physics.

M.V. Lomonosov Gold Medal, awarded to Soviet and foreign scientists for work in natural and social sciences.**

A.M. Lyapunov Gold Medal, awarded for outstanding work in mathematics and mechanics.

I.I. Mechnikov Gold Medal, awarded to Soviet and foreign scientists for outstanding work in microbiology, epidemiology, zoology, the treatment of infectious diseases and for scientific achievements in the field of biology.

D.I. Mendeleyev Gold Medal, awarded to Soviet scientists for important practical work in chemical science and technology.***

I.P. Pavlov Gold Medal, awarded to Soviet scientists for aggregate work on the development of I. P. Pavlov's teaching.

---

* Competitions for the gold medals and prizes are held once every three years.

** Two of each awarded annually.

*** Awarded once every two years at a joint Presidium-level meeting of the USSR Academy of Sciences and the Mendeleyev All-Union Chemical Society.

A.S. Popov Gold Medal, awarded to Soviet and foreign scientists for inventions in the field of radio.

D.N. Pryanishnikov Gold Medal, awarded to Soviet scientists for work on plant nutrition and the use of fertilizers.

A.A. Raspletin Gold Medal, awarded to Soviet scientists for work on radio engineering control systems.*

K.E. Tsiolkovskiy Gold Medal, awarded to Soviet and foreign scientists for work on interplanetary flight.

S.I. Vavilov Gold Medal, awarded to Soviet scientists for outstanding work in physics.

V.I. Vernadskiy Gold Medal, awarded to Soviet scientists for work in geochemistry, biogeochemistry and space chemistry.

A.A. Andronov Prize, awarded to Soviet scientists for work on automatic control theory and for establishing new automation principles.

P.P. Anosov Prize, awarded to Soviet scientists for work in metallurgy, metal science and heat treatment of steel.

A.N. Bakh Prize, awarded to Soviet scientists for work in biochemistry.

V.G. Belinskiy Prize, awarded to Soviet scholars for works of literary criticism and the theory and history of literature.

F.A. Bredikhin Prize, awarded to Soviet scientists for work on astronomy.

A.M. Butlerov Prize, awarded to Soviet scientists for work on organic chemistry.

K.M. Bykov Prize, awarded to Soviet scientists for work in physiology, cortico-visceral physiology,

---

* Awarded once every four years.

physiology of digestion and balneology.

S.A. Chaplygin Prize, awarded to Soviet scientists for original theoretical research on mechanics.

P.L. Chebyshev Prize, awarded to Soviet scientists for works on mathematics.

N.G. Chernyshevskiy Prize, awarded to Soviet researchers for work in the social sciences.

N.A. Dobrolyubov Prize, awarded to Soviet scholars for work on literary criticism.

V.V. Dokuchayev Prize, awarded to Soviet scientists for research in soil science and petroleum geology.

A.Ye. Fersman Prize, awarded to Soviet scientists for work on mineralogy and geochemistry.

Ye.S. Fyodorov Prize, awarded to Soviet scientists for research on crystallography.

D.D. Grekov Prize, awarded for research on the history of the territory of the USSR.

I.M. Gubkin Prize, awarded to Soviet scientists for work on petroleum geology.

A.P. Karpinskiy Prize, awarded to Soviet scientists for work in geology, paleontology, petrography and minerals.

V.G. Khlopin Prize, awarded to Soviet scientists for work on radiochemistry.

V.L. Komarov Prize, awarded to Soviet scientists for research on the botany, taxonomy, anatomy and morphology of plants, botanical geography and paleobotany.

A.O. Kovalevskiy Prize, awarded to Soviet scientists for research on the general, descriptive and experimental embryology of invertebrates and vertebrates.

A.N. Krylov Prize, awarded to Soviet scientists for

work on the theory and application of computer engineering.

G.M. Krzhizhanovskiy Prize, awarded to Soviet scientists for work in power engineering.

N.S. Kurnakov Prize, awarded to Soviet scientists for work in inorganic chemistry, physicochemical analysis and its applications.

L.D. Landau Prize, awarded to Soviet scientists for work in theoretical physics, including nuclear physics and the physics of elementary particles.

S.V. Lebedev Prize, awarded to Soviet scientists for research on the chemistry and technology of synthetic rubber and other high-molecular compounds.

N.I. Lobachevskiy Prize, awarded to Soviet and foreign mathematicians for work on geometry, particularly non-Euclidean geometry.

M.V. Lomonosov Prize, awarded to Soviet scientists for research and discoveries in the field of physics.

A.A. Markov Prize, awarded to Soviet scientists for works on mathematics.

I.I. Mechnikov Prize, awarded to Soviet scientists for work in microbiology, immunology, epidemiology, zoology, the treatment of infectious diseases and important achievements in the field of biology.

D.I. Mendeleyev Prize, awarded to Soviet scientists for original theoretical research in chemistry and chemical technology.

N.N. Miklukho - Maklay Prize, awarded to Soviet researchers for work on the general ethnography of Oceania and Southeast Asia, ethnic anthropology and the geography of the Pacific countries.

V.A. Obruchev Prize, awarded to Soviet specialists for works on the geology and geography of Asia.

L.A. Orbeli Prize, awarded to Soviet scientists for work in the physiology of the vegetative nervous system.

I.P. Pavlov Prize, awarded to Soviet scientists for the best works in physiology.

I.I. Polzunov Prize, awarded to Soviet scientists for research on thermal engineering and new thermal engineering structures.

A.S. Popov Prize, awarded to Soviet scientists for work in radio engineering electronics.

A.S. Pushkin Prize, awarded to Soviet scholars for work on the Russian language and literature.

D.S. Rozhdestvenskiy Prize, awarded to Soviet scientists for work in theoretical and applied optics.

F.P. Savarenskiy Prize, awarded to Soviet scientists for work in hydrogeology.

I.M. Sechenov Prize, awarded to Soviet scientists for experimental and theoretical research in general physiology.

O.Yu. Shmidt Prize, awarded to Soviet scientists for work in geophysics.

K.A. Timiryazev Prize, awarded to Soviet scientists for work on the physiology of plants and general biology.

N.I. Vavilov Prize, awarded to Soviet scientists for research in genetics, selection and plant-breeding.

V.I. Vernadskiy Prize, awarded to Soviet scientists for research on biogeochemistry, geochemistry and space chemistry.

V.P. Volgin Prize, awarded for work in general history and the history of socialist theory.

P.N. Yablochnikov Prize, awarded to Soviet scientists for electrical engineering research and new devices.

N.D. Zelinskiy Prize, awarded to Soviet scientists for research in organic chemistry and petroleum chemistry.

Data Received After Press Deadline

On 24 Nov 1970 the General Assembly of the USSR Academy of Sciences elected the following scientists as full resp. corresponding members of the Academy:

Full members

Vladimirov, V.S.	Dept of Mathematics
Kadomtsev, B.B. Kobzaryov, Yu.B. Lifshits, I.M.	Dept of Gen Physics and Astronomy
Neyman, L.R. Tuchkevich, V.M.	Dept of Physicotech Power Eng Problems
Isanin, N.N. Voronov, A.A. Yanenko, N.N.	Dept of Mechanics and Control Processes
Cherenkov, P.A. Skrinskiy, A.N.	Dept of Nuclear Physics
Kolotyrkin, Ya.M. Postovskiy, I.Ya.	Dept of Gen and Tech Chemistry
Novosyolova, A.V. Sadovskiy, V.D.	Dept of Physicochemistry and Inorganic Materials Technology
Bayev, A.A. Ovchinnikov, Yu.A. Spirin, A.S.	Dept of Biochemistry, Biophysics and Active Physiological Compounds Chemistry
Livanov, M.N.	Dept of Physiology
Shvarts, S.S.	Dept of Gen Biology
Chukhrov, F.V. Kosygin, Yu.A. Kuznetsov, V.A. Shilo, N.A.	Dept of Geology, Geophysics and Geochemistry
Markov, K.K. Obukhov, A.M.	Dept of Oceanology, Atmospheric Physics and Geography

Piotrovskiy, B.B.   Dept of History

Yefimov, A.N.   Dept of Economics

Likhachyov, D.S.   Dept of Lit and Language

Corresp members

Ivanov, V.K.   Dept of Mathematics
Leont'yev, A.F.
Yershov, A.P.
Yershov, Yu.L.
Zolotov, Ye.V.

Abdullayev, G.M.   Dept of Gen Physics and
Amirkhanov, Kh.I.   Astronomy
Denisyuk, Yu.N.
Kagan, Yu.M.
Migulin, V.V.
Nesterikhin, Yu.Ye.
Sharvin, Yu.V.
Shur, Ya.Sh.
Smolenskiy, G.A.
Troitskiy, V.S.
Zuyev, V.Ye.

Fradkin, Ye.S.   Dept of Nuclear Physics
Lobashyov, V.M.
Prokoshkin, Yu.D.

Zhimerin, D.G.   Dept of Physicotech Power Eng
                       Problems

Babakin, G.N.   Dept of Mechanics and Control
Malmeyster, A.K.   Processes
Rumyantsev, V.V.
Vasil'yev, O.F.
Vorovich, I.I.
Yemel'yanov, S.V.

Berezin, I.V.   Dept of Gen and Tech Chemistry
Karpachev, S.V.
Rafikov, S.R.
Voronkov, M.G.
Yevstratov, V.F.
Zhdanov, Yu.A.

Gagarinskiy, Yu.V. Gel'd, P.V. Nikolayev, G.A. Sakharov B.A. Tumanov, A.T. Zolotov, Yu.A.	Dept of Physicochemistry and Inorganic Materials Technology
Georgiyev, G.P. Nichiporovich, A.A. Reymers, F.E. Yelyakov, G.B.	Dept of Biochemistry, Biophysics and Active Physiological Compounds Chemistry
Bekhteryova, N.P. Karamyan, A.I.	Dept of Physiology
Fyodorov, A.A. Galaziy, G.I. Kolesnikov, B.P. Kontrimavichus, V.L. Neunylov, B.A. P'yavchenko, N.I. Sokolov, V.Ye.	Dept of Gen Biology
Bulashevich, Yu.P. Fedotov, S.A. Ivanov, S.N. Krasnyy, L.I. Radkevich, Ye.A. Shipulin, F.K. Vassoyevich, N.B.	Dept of Geology, Geophysics and Geochemistry
Bogorodskiy, V.V. Kalinin, G.P. Kapitsa, A.P.	Dept of Oceanology, Atmospheric Physics and Geography
Krushanov, A.I. Volobuyev, P.V.	Dept of History
Arbatov, G.A. Bunich, P.G.	Dept of Economics
Budagov, R.A. Reizov, B.G.	Dept of Lit and Language
Gvishiyani, D.M. Kopnin, P.V. Rutkevich, M.N. Svechnikov, G.A.	Dept of Philosophy and Law

In 1970 the USSR Acad of Sci established the following institutions:

Institution	Director	
Institute of Atmospheric Optics	Zuyev, V.Ye.	1970-
Institute of High Power Physics	Logunov, A.A.	1970-

In 1970 the following already existing Far Eastern institutions were united into the USSR Academy of Sciences' Far Eastern Scientific Center: Northeastern Inst of Comprehensive Research, Far Eastern Inst of Geology, Inst of Biologically Active Materials, Inst of Soil Biology, Sakhalin Inst of Comprehensive Research, Inst of Vulcanology, Inst of Sea Biology, Khabarovsk Inst of Comprehensive Research.

Institute of Sea Biology, USSR Acad of Sci's Far Eastern Scientific Center	Zhirmunskiy, A.V.	1970-
Kazan' Institute of Biology	Gusev, N.A.	1970-
	Chairman	
Sci Council for Geomagnetism	Troitskaya, V.A.	1970-
Sci Council for Polymolecular Compounds*	Andrianov, K.A.	1970-
Sci Council for Analytical Chemistry	Alimarin, I.P.	1970-
Sci Council for Chemistry and Technology of Semiconducting and Highly Pure Materials	Novosyolov, A.V.	1970-
Sci Council for Int Relations Concerning Regional Research, Presidium, USSR Acad of Sci	Nekrasov, N.N.	1970-

Sci Council for the Cryology of the Earth	Mel'nikov, P.I.	1970-
Sci Council for the Comprehensive Problem of "The Methods of Direct Transformation of Heat Energy into Electric Power"*	Millionshchikov, M.D.	1970-
Sci Council for the Problem of "The Chemistry of Cellulose and Its Main Components"	Kalnin', A.Ya.	1970-
Sci Council for the Comprehensive Problem of "The Economic Competition of the Two Systems"	Rumyantsev, A.M.	1970-
Sci Council for the Comprehensive Problem of "The Contemporary Problems of Development Countries"	Tyagunenko, V.L.	1970-
Sci Council for the Problem of "The Physicochemical Principles for Obtaining Heatproof Inorganic Materials"	Tananayev, I.V.	1970-
Sci Council for Catalysis*	Boreskov, G.K.	1970-
National Comt for Implementation of the USSR Acad of Sci's Membership in the Int Council of Scientific Unions (National Comt of the Int Council of Scientific Unions, Presidium, USSR Acad of Sci)	Vinogradov, A.P.	1970-

---

* Date of establishment unknown.

SOURCES

1. RESEARCH WORKS AND PERIODICALS

Akademiya nauk SSSR. Kratkiy ocherk istorii i deyatel'nosti (The USSR Academy of Sciences. A Brief Outline of Its History and Work), exec ed: V. A. Vinogradov, "Nauka" Press, Moscow 1968.

Istoriya Akademii nauk SSSR (The History of the USSR Academy of Sciences), chief ed: K. V. Ostrovityanov, USSR Acad of Sci "Nauka" Press, 2 vol, Moscow-Leningrad, 1958-1964.

G. A. Knyazev, Kratkiy ocherk istorii Akademii nauk SSSR (A Brief Outline History of the USSR Academy of Sciences), USSR Acad of Sci Press, Moscow-Leningrad, 1945.

G.A. Knyazev and A.V. Kol'tsov, Kratkiy ocherk istorii Akademii nauk SSSR, expanded 2nd and 3rd ed, "Nauka" Press, Moscow-Leningrad, 1957 and 1964.

G.D. Komkov, O.M. Karpenko, B.V. Levshin and L.K. Semyonov, Akademiya nauk SSSR - shtab sovetskoy nauki (The USSR Academy of Sciences - Headquarters of Soviet Science), "Nauka" Press, Moscow, 1968.

Otchyot o deyatel'nosti Rossiyskoy Akademii nauk za 1919 god (Report on the Russian Academy of Sciences' Work for 1919), Petrograd, 1920.

Vestnik Akademii nauk SSSR (USSR Academy of Sciences' Herald), Leningrad-Moscow, 1930-1970.

2. REFERENCE WORKS

Bibliografiya izdaniy Akademii nauk SSSR. Yezhegodnik (A Bibliography of USSR Academy of Sciences' Publications. Annual), ed: S. P. Luppov, USSR Acad of Sci Press, vol 1-11, Leningrad, 1957-1969.

Biograficheskiy slovar' deyateley yestestvoznaniya i tekhniki (Biographical Dictionary of Natural Science and Engineering Workers), exec ed: A. A. Zvorykin,

"Large Soviet Encyclopedia" Press, 2 vol, Moscow, 1958-1959.

Bol'shaya Sovetskaya Entsiklopediya (Large Soviet Encyclopedia), 1st ed, vol 1-65, Moscow, 1926-1947.

Bol'shaya Sovetskaya Entsiklopediya (Large Soviet Encyclopedia), 2nd ed, vol 1-51, Moscow, 1949-1958.

Yezhegodnik Bol'shoy Sovetskoy Entsiklopedii (Large Soviet Encyclopedia Yearbook), Moscow, 1957-1969.

Malaya Sovetskaya Entsiklopediya (Small Soviet Encyclopedia), 2nd ed, vol 1-10, Moscow, 1958-1960.

Sovetskaya Istoricheskaya Entsiklopediya (Soviet Historical Encyclopedia), "Soviet Encyclopedia" Press, vol 1-12, Moscow, 1961-1969.

Ukrains'ka Radyans'ka Entsyklopediya (Soviet Ukrainian Encyclopedia), Ukr Acad of Sci Press, vol 1-17, Kiev, 1959-1965.

Records and archives of the Institute for the Study of the USSR.

Q
127
R9 M42
pt. I

SEP 15 1971